Turkey Red

TEXTILES THAT CHANGED THE WORLD

Series Editor: Linda Welters, University of Rhode Island, USA

ISSN: 1477-6294

Textiles have had a profound impact on the world in a multitude of ways—from the global economy to the practical and aesthetic properties that subtly shape our everyday lives. This exciting new series chronicles the cultural life of individual textiles through sustained, book-length examinations. Pioneering in approach, the series will focus on historical, social and cultural issues and the myriad ways in which textiles ramify meaning. Each book will be devoted to an individual textile or to the dye, such as indigo or madder, that characterizes a particular type of cloth. Books will be handsomely illustrated with color as well as black-and-white photographs.

Previously published in the Series

Tartan, Jonathan Faiers

Felt, Willow Mullins

Cotton, Beverly Lemire

Digital Textile Printing, Susan Carden

Tweed, Fiona Anderson

Turkey Red

Julie Wertz

BLOOMSBURY VISUAL ARTS
LONDON • NEW YORK • OXFORD • NEW DELHI • SYDNEY

BLOOMSBURY VISUAL ARTS
Bloomsbury Publishing Plc
50 Bedford Square, London, WC1B 3DP, UK
1385 Broadway, New York, NY 10018, USA
29 Earlsfort Terrace, Dublin 2, Ireland

BLOOMSBURY, BLOOMSBURY VISUAL ARTS and the Diana logo are trademarks of
Bloomsbury Publishing Plc

First published in Great Britain 2024

Cover design by Adriana Brioso

A catalogue record for this book is available from the British Library.

A catalog record for this book is available from the Library of Congress.

ISBN: HB: 978-1-3502-1651-8
        PB: 978-1-3502-1650-1
        ePDF: 978-1-3502-1652-5
        eBook: 978-1-3502-1653-2

Typeset by RefineCatch Limited, Bungay, Suffolk
Printed and bound in India

To find out more about our authors and books visit www.bloomsbury.com
and sign up for our newsletters.

This book is dedicated to everyone whose labor is a part of the Turkey red story.

Contents

Preface

Most of the material for this book comes from research conducted at the University of Glasgow for my doctoral thesis, which began as an investigation on the use of early synthetic dyes in late nineteenth-century Scotland and evolved into a focused study on the process and chemistry of Turkey red dyeing and textiles. It was intended to be a starting point for the project, but as I conducted my literature review I became increasingly fixated on the persistent questions about its materiality and origins. With the support of my supervisors, I focused my research on Turkey red, primarily within the context of late nineteenth-century Scotland. The project was a multidisciplinary investigation that considered the history, chemistry, and process of Turkey red dyeing. I gave many public talks on Turkey red to audiences around Glasgow and across the UK, and was continually delighted and impressed with the interest it generated.

My intent with this book was to expand the scope beyond Scotland, and discuss more about the origins, use, and object record. I began writing the proposal in the autumn of 2019 at the invitation of Dr. Linda Welters, series editor for *Textiles that Changed the World*, and submitted it just after the start of 2020. Part of my plan was to visit archives and collections in France and Switzerland, and revisit Scottish Turkey red material. The outbreak of the coronavirus pandemic in March 2020 and accompanying travel restrictions effectively canceled these plans. The availability of digitized materials and sources, particularly searchable texts, meant a more comprehensive story of Turkey red than ever before could be told, so it was still possible to write the book with a broad overview of the industry in these places but no new examination from primary source material. The discussion of Turkey red process, materiality, trade, technology, history, industry, and use in this book covers much of the story, and *Andrinople, Le rouge magnifique*[1] and *Colouring the Nation*[2] examine the Mulhouse and Scottish industries, respectively. Still, there remains more to be told about Turkey red, but that is for another text.

Illustrations

Acknowledgments

My first and deepest gratitude goes to Andy, for everything.

A huge thanks to series editor Linda Welters for her suggestion to submit the proposal for this volume. I had not planned to write anything after my dissertation, and I am grateful for the opportunity to have done so. I want to thank Bloomsbury for their interest and support in publishing this book. The Pasold Research Fund kindly provided a grant to obtain publication rights for some of the images included here, and I appreciate their time, consideration, and funding.

The majority of my research on Turkey red was made possible by the generous funding of the Lord Kelvin Adam Smith scholarship scheme at the University of Glasgow, which eventually led to this book and also some wonderful opportunities in research and work. I was supervised by Dr. Anita Quye and Dr. David France, to whom I am incredibly grateful for such a life-changing opportunity. I also want to thank my mom for encouraging and supporting a lifelong interest in making things.

The Glasgow University Archives and Special Collections were an invaluable resource for this research. I would like to thank everyone who helped me search the collections and retrieve archive material, in particular retired University Archivist and Deputy Director of the Library Lesley Richmond and GUASC colleagues Ela Gorska-Wiklo, Bob MacLean, and Claire Daniels. Other collections access was generously provided by Susan North and Jo Hackett at the Victoria and Albert Museum.

I would not have done any of this if my friend, Dr. Jennifer Brand, had not suggested I apply to Glasgow for the original project. I am grateful to Dr. Sally Tuckett for sharing her knowledge of Turkey red prints, and for her support of my application to the Pasold. Thanks also go to Mary McWilliams for acting as a reference. Much of the discourse in this book was refined through talks on Turkey red that I was invited to give, and I appreciate all the interest and engagement from attendees over the years. I always enjoy talking textiles with my friend Laura Garcia Vedrenne, who has so helpfully answered my many questions.

All translations are by the author unless otherwise noted.

Introduction

The story of Turkey red textiles is one of innovation, trade, knowledge exchange, industrialization, and globalization. It spans both centuries and continents, appearing in an array of objects from many cultures around the world. Deemed "the most complicated and tedious operation in the art of dyeing," the Turkey red process created "reds possessing great brilliancy and beauty" on cotton (or occasionally linen) fibers, which was impossible by other means until the late nineteenth century.[1] Its name (sometimes as Turkish red) comes from the perspective of Western European buyers who imported it from the Eastern Mediterranean before the mid-eighteenth century. It was also called Adrianople red after a place in northwest Turkey (present-day Edirne) where it was dyed in large quantities. In French, it is called *rouge turc* and *rouge d'Andrinople*. Another French name, *rouge des Indes*, implies earlier origins in India. In German, its name is *Türkischrot* and in Dutch it is *turksrood*. The color has no direct association with the distinctive North American red-wattled birds, and although it is the same color as the Turkish national flag, there is no evidence linking it to the textile.

Although Turkey red has gradually disappeared from daily life over the last century, its allure is strong enough that it has not been entirely forgotten. A 1995 exhibition at the Musée de l'Impression sur Étoffes in Mulhouse explored the history of Turkey red in France using the museum's collection of nineteenth-century textiles.[2] The TextielMuseum in Tilburg had a 2013 exhibition called *Turkish red & more*, a collaboration with design studio Formafantasma to showcase Turkey red and some of the archival material held by the museum about Dutch dyers.[3] Around the same time, a project at the National Museums Scotland called *Colouring the Nation* examined their archive of nineteenth-century printed Scottish Turkey red.[4] Turkey red, which often appears in quilt blocks from this time, has been the focus of multiple events at the International Quilt Museum at the University of Nebraska-Lincoln, and has been the subject of practical craft dyeing courses for those interested in the process.[5]

Most of the research for this book comes from my doctoral thesis at the University of Glasgow (2012–16), which examined the process and chemistry of Turkey red through re-creation and analysis.[6] This book expands on the topic and attempts to answer more about the pre-industrial origins of Turkey red and how it came to Western Europe through material and knowledge exchanges. It follows the history of Turkey red textiles through a process-based study, tracing possible origins through similar practice on a material level by considering the chemistry of the individual treatments necessary to build the color complex. An interdisciplinary examination of Turkey red from both historical/archival and material/scientific perspectives enables a more comprehensive understanding of the topic. Growing specialization in research fields throughout the last century have increasingly separated the humanities and sciences, generating new sub-disciplines as we learn about our world. This has enabled more focused, deep investigations, which are of great benefit to knowledge, but the cost of being more versed in one subject is being less so in others. The emergence of modern scientific thought following the Renaissance and the subsequent Age of Enlightenment in Europe produced many great minds whose contributions to humanity encompass subjects that may seem disparate to us today. Much of the historical knowledge about Turkey red came from these polymaths whose work would be considered interdisciplinary today, but at the time would have been seen as a practical application of available knowledge to answer new questions.

The term Turkey red can refer to many things. Chapter 1 discusses some terminology and context for the work, including what is not considered Turkey red here, and the historical reputation of Turkey red. It gives a definition of Turkey red in material and chemical terms, and describes identifying characteristics of Turkey red textiles, which informs how surviving objects and documents are interpreted in this work. This lays the foundation for Chapter 2, which uses the given definition of Turkey red as reference to compare with historical dyeing practice in places like India, where it likely originated. Previous histories tend to focus on the European industry in the nineteenth century, for which there is the most surviving documentation. The chapter also discusses some of the history and use of madder root in the context of Turkey red dyeing, and the subsequent development of madder-derived dye products and synthetic alizarin. This remarkable feat of early synthetic chemistry was driven by the demand for colorant from the Turkey red industry, and revolutionized the process without changing it on a material level.

The distinctive Turkey red process, its ingredients, and chemistry are discussed in Chapter 3. It is presented stepwise following the essential sequence of treatments that yields the remarkable red cotton, along with regional and historical variations identified in the previous chapter. The technological transitions of industrialization, chemical manufacturing, and their impact on Turkey red are also examined. Chapter 4 focuses on the printing of Turkey red, another distinctive process that requires selectively discharging the color from fully-dyed cotton before replacing it with another color. This was often done using a distinctive, yet limited palette, though the myriad designs for markets both domestic and international seems limitless. In Chapter 5, an overview of the European Turkey red industry in the nineteenth century is presented. Previous studies of Turkey red have focused on certain major producers, but there is no comprehensive study and for many places there is very little published on the topic. This book identifies centers of production that may have untold histories of local Turkey red production documented in archive records, but telling those stories in full is beyond its scope. Finally, a selection of objects known or suspected to be Turkey red based on the characteristics and definition used in this text is presented in Chapter 6. These objects were chosen for their representativeness of the many ways in which Turkey red was used.

Historical information about Turkey red comes from disparate sources, which are significantly more accessible in the digital era. Digitized books, treatises, newspapers, and other texts enable remote research in library and archive collections, and searchable texts make it possible to scan for keywords in documents and databases. Dyeing manuals or treatises, the texts that function as guidebooks for dyers, provide technical insight into the specifics of the process to dye it. Some discuss the chemistry or analytical work, which also appear in scientific publications. Trade publications touting products or recent advances in manufacturing are also relevant and can come from a manufacturing perspective or from sales, like retail catalogs. Social histories describe other aspects of Turkey red, and newspapers provide information about when and where it was being made and sold. Although information is more accessible than ever before, it still depends on what historical material was preserved. Mass-printed publications are inherently more likely to survive into the present day. Documents can be created with the necessary materials (paper and ink) and a literate author, whose language may not be known to future readers. Manuscripts, letters, travel accounts, and recipe notebooks from those with firsthand experience are a valuable resource, but many archive collections are not digitized and catalog notes may be brief. These objects also have a lower rate of preservation. As such, there is doubtless still more to be discovered about Turkey red, and also some things that may be lost to time.

An interdisciplinary approach is crucial for a full understanding of what Turkey red is in terms of its material and technological significance and its continued legacy as a physical and intangible historical artifact. Every effort is made to provide a full contextual explanation of the concepts presented so that the reader, regardless of background, will find the text accessible. Some explanations may seem redundant, but the intent is to reach as broad of an audience as possible and to do justice to a complicated topic. For ease of reading, chapter note references only refer to source citations and do not contain any commentary that requires skipping forward in the text. Maps are drawn with modern-day borders. Approximate modern values for historical prices were calculated using resources like The National Archives Currency Converter and the Official Data Foundation.[7] Objects discussed that are not included as plates should be possible to view in the online collections database for the holding institution using the accession number provided.

A glossary of terms is included with chemical formulas for compounds mentioned in the text. Chemical terminology has evolved since the first publications about Turkey red and its chemistry appeared in the late eighteenth century, and there are also regional differences in language, e.g., "aluminium" and "sulphur" in Commonwealth English versus "aluminum" and "sulfur" in US English. The International Union of Pure and Applied Chemistry (IUPAC) defines standard terminology for international use, which is the convention that is followed in this book with respect to compound and material names.

1

The Most Brilliant Color Dyed on Cotton

In the broadest sense, the term Turkey red refers to a color around true red, though hue is always subject to individual perception. This book is about the textiles, however, which were made following the eponymous process. Figure 1.1 shows samples of plain Turkey red fabric from the late nineteenth century in a characteristic shade. Upon close examination, subtle variations in the hue of historical Turkey reds become apparent, so a textile can be authentic Turkey red without being one exact shade as long as it has been dyed according to the specified method. The red can be produced on wool or silk fibers by other methods, but this is only the same color as Turkey red and not a true Turkey red, which is defined in this chapter. Today, it is even possible to purchase a jar of synthetic fiber-reactive dye, which have been commercially available since the mid-1950s, in a shade called Turkey red. Again, the fabric would only be Turkey red in color, even if it were applied to cotton or linen. Another point of confusion is Turkey carpets, which are typically woolen and thus not true Turkey red, although they may be the same hue.

Turkey red was widely reputed for its particularly vivid color and for its fastness, resisting fading from light exposure, repeated washing, rubbing, and bleaching. These features made it particularly useful for some applications, but limited in others unless a very strong color was required. This chapter defines Turkey red as a process using particular materials and treatments, presenting a framework to interpret the history and object record later in the book. It also discusses the historical reputation of Turkey red, and the criteria for assuming whether something is Turkey red when interpreting historical objects and documentary records. While an analytical identification of Turkey red in objects would always be the most robust, it is acknowledged that the costs are often

Figure 1.1 Plain Turkey red samples, late nineteenth century. *University of Glasgow Archives and Special Collections, Records of United Turkey Red Co Ltd, GB248. UGD 13/8/9*

too great for most institutions to bear and this book aims, as much as possible, to provide guidance for seeking Turkey red in collections through more accessible means.

Defining Turkey Red

A true Turkey red is dyed on cotton or, rarely, linen, that was prepared with an oil treatment, then an aluminium mordant, and dyed in a bath with madder or synthetic alizarin containing calcium. The order of these treatments should not vary, but the origin of the materials and the conditions of individual steps may. As with any dyeing process, it is essential to start with scoured fibers. On a chemical level, a Turkey red complex forms when fatty acids from the oil treatment attach to the cellulose of the cotton fibers, which subsequently attract aluminium atoms. These complex with the dye molecules and calcium, coloring the fiber (discussed further in Chapter 3). The resulting color should be a vivid, primary red, though with manipulation (or contamination) it can range from pink to maroon. For the most part, historical Turkey red is a pure red shade. Due to the particulars of the process, it is not possible to have a single-application Turkey red as a liquid or powder, or for it to exist independently from the fiber onto which it is dyed. Rather, it is "built" on cotton through the oil, aluminium, and dyeing steps. Occasionally, texts mention "Alizarin red," a shorter process sometimes made alongside Turkey red. It started with aluminium, then proceeded to dyeing before finishing with an oil treatment. Alizarin reds were not as brilliant or as fast as Turkey red, but were still considered to be good and an acceptable, cheaper alternative for it.[1] The difference in quality was attributed to the change in oil application, which must always be first in Turkey red.[2]

The definition used here encompasses product variations by color, process, and provenance, including materially similar textiles dyed in different centuries and locations using a range of ingredients that fulfill the same function. Some methods contain additional steps, like treatments with ruminant dung in the oil bath or animal blood in the dye bath, or an application of tannins between the oiling and aluminium. These function as auxiliary agents or assistants to the process, rather than essential elements, and it is possible to dye Turkey red without them. Advances in chemistry and manufacturing enabled two significant changes to industrial Turkey red production in the late nineteenth century. Both involved replacing a natural ingredient with a manufactured one, and represented significant developments for the field of textile dyeing as a whole. Oil treatments were usually done with rancid plant oils until the 1870s, when the development of Turkey red oil, a modified castor oil, significantly reduced the length of the process. The red color originally came from a range of plant sources, most significantly madder roots from the species *Rubia tinctorum*. These natural sources contain anthraquinone dyes, with alizarin as the major component in most. The textile industry began to adopt synthetic dyes in the second half of the nineteenth century, after William Henry Perkin invented mauveine in 1856. In the late 1860s, commercial synthetic alizarin quickly replaced madder in Turkey red. Madder and synthetic alizarin are discussed further in Chapter 2.

By the late nineteenth century, chemists had begun to develop single-application synthetic red dyes with good stability, though viable products for cotton remained elusive. The first plausible candidate was Congo Red in 1884, which resembled Turkey red when freshly dyed and was sometimes sold in packages labeled as such for home use (see Figure 1.2). Unfortunately, it had extremely poor fastness and easily fades to pink or brown.[3] It is unclear to what extent consumers understood these products dyed a Turkey red *color* and not genuine Turkey red, though after use

Figure 1.2 An early twentieth-century packet of Putnam Fadeless Dyes in Turkey Red. This produced the desired hue but was an early synthetic dye that would fade quickly. *Julie Wertz*

the difference would quickly be obvious. The next synthetic reds were alpha-naphthylamine red and Naphthol AS dyes, which were better but not competitive.[4] In the early twentieth century, synthetic red azo dyes with acceptable hue and fastness, which could be applied directly to cotton without a lengthy preparatory process, began to supersede Turkey red dyeing. It went into decline and ended as a commercial venture in the mid-1930s.[5]

Although it has vanished from our collective consciousness, in the nineteenth century Turkey red was sufficiently well-known by name, due to its color and reputation for durability, that it could be mentioned without explanation in publications intended for a general audience. An article titled "Sunday School Object Lesson" published in the Christian Recorder on August 15, 1868, instructs teachers to describe the "process of creating red dyes, and the general permanence of their colors." It uses the fastness of Turkey red, and its resistance to bleaching, as an allegory for the accumulation of sin on one's soul. The teacher was to ask the students whether a piece of Turkey red can ever truly become white—the answer here being (incorrectly, as discussed in Chapter 4) "no"—and that the only way to cleanse oneself of sin was the blood of Jesus Christ, creating an unintentional analogy between salvation and acidified bleach.[6] Turkey red was mentioned in popular publications like Godey's Lady's Book, a nineteenth-century American magazine for women that included short stories, poetry, articles, sheet music, and engravings with fashion plates of the latest styles. It became a household name in the United States in part due to the settlement of Scottish immigrants throughout North America during the nineteenth century, who brought Turkey red textiles from home and became the foundation for an export market. It was also associated with American cultural icons like the cowboy wearing a red bandana, a commonplace item made of Turkey red.[7]

A Reputable Red

Turkey red is frequently described in the literature as brilliant, vivid, bright, and fiery. Although commonplace today, at the time a stable, true red shade on cotton was a special thing. It has an intensity of tone and liveliness not obtained by other madder dyeing methods, and was visibly distinct from other red dyes on cotton—indeed, "one of the best and handsomest colors" available.[8] The color was saturated enough that, with the exception of black, printing over it with another color was a wasted effort.[9]

It was acknowledged that the most beautiful reds came from using fatty material in the preparatory process, and madder dyeing without the oil or fat, sometimes called "madder red" or *rouge de garance*, would not be as bright or colorful and more like a shade of puce.[10] A late nineteenth-century experiment found that dyeing without oil resulted in an uneven, dirty purple color.[11] The object record shows that skilled printers of madder reds could obtain an attractive hue, though not necessarily as fiery as that of Turkey red, which cannot be made by direct printing. The patterns were printed in red over a white ground by applying aluminium mordant to the fabric with blocks, plates, or rollers, typically with detailed engraved patterns. The mordant attracted dye in the madder bath, turning the printed area red. These textiles, shown in Figure 1.3, are often referred to as *toile de Jouy*. An eighteenth-century engraving of a madder dyeing operation is shown in Figure 1.4.

Figure 1.3 Eighteenth-century copper plate print on cotton, probably with madder. These textiles are referred to as "madder reds" in technical literature and were not made following a Turkey red process. *Cooper Hewitt, Smithsonian Design Museum. Bequest of Elinor Merrell. 1995-50-129-a,b*

Figure 1.4 Print depicting a dyeing workshop for "ordinary colors" like madder, 1780. *L'art du fabricant de velours de coton précédé d'une dissertation sur la nature, le choix et la préparation des matières et suivi d'un traité de la teinture et de l'impression des étoffes de ces matières, par M. Roland de La Platière, . . . Roland de La Platière, Jean-Marie. impr. de Moutard. Paris. 1780. Source gallica.bnf.fr / BnF*

A key aspect of Turkey red was its reputation for fastness to light and washing. The cotton fibers had to be of sufficient quality to withstand the arduous dyeing process, and Turkey red fabrics gained a reputation for being "absolutely indestructible."[12] The historical attestations are numerous and often superlative, ranging from "absolute" to "unrivalled," and were somewhat surprising to the uninitiated.[13] Dyeing texts said Turkey red should resist fading when placed in *aqua fortis* (dilute nitric acid) for fifteen minutes. It was also more resistant to alkalis, soaps, and acids in general, and sold well in sunny climates like India and Southeast Asia where light fading was a greater concern.[14] Systematic evaluation using mid-twentieth-century industrial standardized testing determined Turkey red has a middling to high fastness.[15] Although remarkably robust to light, it will eventually succumb to extended light exposure. Figure 1.5 shows part of a handkerchief that received significant light damage prior to being collected and the contrasting unfaded areas that were protected by a label which has since partially detached.

Another recognized feature of Turkey red was its distinct resistance to bleach (aqueous calcium or sodium hypochlorite) which was exploited in two ways. In normal bleaching conditions, which have an alkaline pH, Turkey red resists losing color where other textiles quickly blanch and fade. Affordable white and red cotton checks, popular in the Caribbean and in Africa, were made by weaving Turkey red and unbleached cotton yarns, then bleaching the finished textile to brighten the white while the red remained unchanged.[16] A late eighteenth-century painting of a market in Dominica (Figure 1.6) depicts women wearing red and white checked shawls with a red border made in this fashion. This style of textile was also suitable for objects that might be washed or bleached repeatedly, like tablecloths and kitchen towels, which were often woven with plain linen and Turkey red yarn.[17] If, however, a weak acid is added to the bleach, Turkey red rapidly returns to white.[18] This process was exploited in the discharge printing process, discussed further in Chapter 4.

A Complicated Process

Historically, the means to dye Turkey red were mysterious.[19] It was described as "a most intricate and troublesome process" and "the most complicated and tedious operation in the art of dyeing."[20] Western Europeans invested substantial money and effort to acquire practical knowledge of Turkey red. Some eighteenth-century publications said that instructions communicated by someone who had only observed the dyeing were insufficient to replicate it, either due to inaccuracy, intentional concealment of some detail, or circumstances unknown and not communicated by the author. Texts state that close observation alone was usually not enough to learn it, and that those who were successful tended to keep their secrets.[21] Most knowledge seems to have transferred practically, through funded representatives sent abroad to learn the process and migrant dyers sharing their own experience.[22] More about the dissemination of Turkey red is discussed in Chapter 2.

Despite this, there is evidence it was possible to successfully dye Turkey red by following a written method. This probably depended on how carefully one followed the directions, the availability of the correct ingredients, and the dyer's practical skills. An 1811 American article describes the author's wife attempting to dye Turkey red following a recipe in *The Domestic*

Figure 1.5 A close image of a handkerchief bearing the label A. Orosdi, Paris. The lower and left edges where the label has broken away reveal the original bright red below. The handkerchief was made for the Tunisian market and was probably left in a sunny shop window before being collected. *Philadelphia Museum of Art. Gift of the Philadelphia Commercial Museum (also known as the Philadelphia Civic Center Museum), Philadelphia, Pennsylvania, 2004. 2004-111-75*

Encyclopedia, resulting in a beautiful red color that she wove into a cloth which was regularly worn and washed, suffering no loss of color.[23] Some of the difficulty probably came from attempted shortcuts, and success depended on controlling many uncertain factors.[24] Key among these was the need for quality ingredients, some of which inherently varied as natural products and could be contaminated with adulterants.[25] A late eighteenth-century account of Armenian dyers on the

Figure 1.6 Painting of the linen market in Dominica by Agostino Brunias, late eighteenth century. The red stripes of the skirts, shawls, and turbans worn by the women were made with Turkey red yarn. *Linen Market, Dominca. Agostino Brunias. Yale Center for British Art, Paul Mellon Collection. B1981.25.76*

Volga Delta said that their Turkey red was "sometimes brighter and sometimes darker," allegedly due to their not paying attention to ingredient proportions and a fluctuating quality in the madder supply.[26]

A repeated assertion in the literature is that Turkey red could only be dyed on yarn until around 1810, when Daniel Koechlin of Mulhouse, France, successfully found a means to dye cloth.[27] The exact nature of his innovation is unclear when comparing published methods from before and after this date. It is difficult, in such a long and complex process, to obtain an even (the technical term is level) color on fabric. A late nineteenth-century expert in Turkey red said that the only difference between cloth and yarn dyeing was the mechanics of the application, which explains why publications do not indicate the specifics of Koechlin's contribution—none of them give detailed handling directions.[28] A possible factor in the emergence of cloth dyeing around this time is the mechanization of cotton spinning in the late eighteenth century, since the greater tensile strength of machine-spun yarn made it less susceptible to chafing and distortion during the numerous treatments.[29]

Turkey red dyeing was "very long, expensive and difficult," and "only profitable when produced in large quantities," effectively a specialty trade with a relatively high investment and operating cost.

One early publication advised that "a dyer of Turkey Red yarn must have his dye-house calculated for that purpose only," limiting the owner's ability to sell other products. Not every establishment was capable of turning out "first-class products," and even reputable dyers expected a certain percentage of failed batches due to the numerous factors involved.[30] A 1786 estimate from Lancashire determined that dyeing cost four shillings per pound of cotton, roughly £16 in twenty-first-century currency. An 1815 American source estimated it would still be profitable to make Turkey red even if it cost two dollars per pound (roughly $40 in twenty-first-century currency) because so little yarn by weight was needed to weave a shawl. Later in the nineteenth century, it was "reputed to be the most profitable of all the cotton finishing sectors," showing that despite the associated costs and difficulty Turkey red could be a worthwhile venture.[31] Part of the burden was the intensive amount of washing and drying that occurred between each treatment, which required a large water supply, and a lot of labor and time.

As such, the desire to shorten or economize on the process was always present, though only so much could be done without compromising on a genuine Turkey red process.[32] Some, but not all of its chemistry was understood, which reduced failures and facilitated optimizing the treatments. Dyers knew that iron contamination in the aluminium mordant, or from fittings or tools, affected the color, creating "a brick-dust shade" and purple spots.[33] Despite knowing oil was essential, dyers and chemists could not fully explain its role using nineteenth-century chemistry.[34] This hindered, but did not discourage research on Turkey red, and publications appear as early as the 1770s on the systematic testing of various solvents and bleaching agents.[35] Although many eminent dyers and scholars studied and experimented throughout the nineteenth and into the twentieth centuries, by the time chemistry advanced enough to answer the remaining questions production had gone into decline and research shifted toward optimizing new techniques.[36]

Identifying Turkey Red

When is it reasonable to assume red cotton is Turkey red? Limited analytical resources and the number of objects in collections often make confirmation by material identification impossible. Physical and contextual evidence can indicate whether something is Turkey red, though any designation based on reasoning without material analytical confirmation should always be acknowledged. Turkey red was called "the only method by which cotton goods could be dyed a bright red of reasonable fastness at moderate cost."[37] Methods in dyeing treatises from the eighteenth and nineteenth centuries support this assertion. Other natural red dyes, like barwood (*Pterocarpus erinaceus* or *soyauxii*), were sufficiently close to Turkey red in color, but with a significantly lower light and wash fastness.[38] It is not clear to what extent these were sold. Cochineal (*Dactylopius coccus*), a scale insect native to the tropics and subtropics of the Americas, was also valued for the vivid reds it produces, but the color was significantly better on wool and silk. On cotton and linen it gave "dull, lifeless reds not worth its cost."[39] In 1779, a substantial reward of £5,000 (£430,000 in twenty-first-century currency) was given by the British government to a Mr. Berkenhout for experiments with cochineal on cotton, but nineteenth-century dyeing texts still state it was unsatisfactory for this fiber.[40] It is unclear whether Central American textiles that contain vivid red cotton, like a *huipil* in the Dallas Museum of Art collection from the Ixtil culture of Guatemala (cat. no. 2008.223), contain cochineal due to the lack of technical studies on these objects.[41] It is possible local dyers with generations of experience using cochineal were able to obtain better colors on cotton than European practitioners, but for objects not from Central American cultures there is very little indication cochineal is common on red cotton. Another textile technique materially similar to Turkey red is *kalamkari*, which involves preparing the cotton with fat (often buffalo milk) and printing an aluminium mordant.[42] There is no indication in the literature, however, of any confusion between Turkey red and *kalamkari*, the latter of which likely has a lower fastness due to the manner of application.

The function of an object can inform whether an object is Turkey red or a cheaper alternative. Since wash fastness was a key characteristic, objects likely to be frequently laundered (e.g. linens, bedding, high-use garments like aprons, children's clothing, etc.) more often featured Turkey red yarn or cloth. Recent examinations of Turkey red in nineteenth-century quilts suggests the rigorousness of the dyeing process makes it more susceptible over time to wearing thin from abrasion and losing color in high-relief areas, while enduring better in low places near quilting lines and seams. The wear pattern, where it becomes streaked with white rather than uniformly fading, also indicates that the color is concentrated near the surface.[43] This is consistent with its reputation for being less fast (though still very good) to color transfer by rubbing. The technical term for this, crocking, is also what can happen when new, dark blue denim rubs against pale upholstery. Worn Turkey red twill can sometimes take on the appearance of wool and may be mistaken as such with only casual visual inspection.[44] The crocking and striated wear patterns indicated to nineteenth-century scholars that the chemistry of Turkey red was different from that of other red-dyed textiles.[45] Bandanas, or handkerchiefs, were also consistently made of Turkey red. Structurally robust pile fabrics like velvet and flannelette are less likely to withstand the rigorous dyeing process, and less likely to be Turkey red. Objects made after 1884 may contain synthetic reds, though other features can still help inform an identification. Prints, with the distinctive colorway, are easier to identify than plain reds, and are further discussed in Chapter 4. Most of the objects discussed in this book have not been confirmed analytically to have Turkey red, and were selected because they were made before 1920, conform to visual descriptions and known Turkey red, and have some additional feature (use, fading pattern, etc.) indicative of Turkey red as discussed here.

Material Record

Turkey red is found in collections around the world, showing the extent to which it was produced and traded, though provenance in catalog records is variable and objects are frequently not identified as or confirmed to be Turkey red. The survival of historical textiles depends on a number of factors—major ones being temperature fluctuations, humidity, light exposure, and social upheaval. Before the Industrial Revolution, the resources and labor required to make a finished garment meant wardrobes for most people were limited in size relative to today. Washing accelerates the breakdown of fibers and colorants, meaning the lifespan of an actively used textile decreases significantly. Both the cotton fiber and the Turkey red complex were more robust to laundering, which made it useful for heavily-used items. Such objects are still less likely to survive, and to be considered worthy of preservation. Collections are subject to individual and institutional preferences. Generally, cotton objects were considered to have a lower value, making them less likely to be collected.[46] Still, enough textiles have survived to provide a fascinating and colorful visual history, though no doubt an incomplete one due to the vagaries of preservation.

The object record is evidence of how and by whom it was used at various times. The Turkey red found in collections is overwhelmingly from the second half of the nineteenth century, the peak of demand and production. The industrial practice of building albums of swatches for internal records, and sometimes for advertising to customers, preserved thousands of samples now found in collections and archives. Examples of known Turkey red from around the world are found in the collections of the Musée de l'impression sur Étoffes (Mulhouse, France), the National Archives (London, England), the Science and Industry Museum (Manchester, England), the Victoria and Albert Museum (London, England), National Museums Scotland (Edinburgh, Scotland), the University of Glasgow Scottish Business Archive (Glasgow, Scotland), as well as many other smaller institutions. More about historical Turkey red use is discussed in Chapter 6.

Conclusion

The story of Turkey red is extensive and complex; to fully tell it requires an interdisciplinary investigation into the history of textile manufacturing, trade, chemistry, and the exchange of knowledge. The textile was as renowned for its fastness to light, washing, and bleach, as it was for its brilliant hue and lively patterns. It was difficult to gain knowledge of Turkey red dyeing, and practice was resource-intensive and notoriously difficult. The definition of Turkey red dyeing as a process where cotton is treated with a fatty oil followed by an application of aluminium, and dyed with anthraquinones like madder or synthetic alizarin in the presence of calcium, the resulting textiles being Turkey red, is used here as a framework to explore the history of the process, its chemistry, and historical use. By defining Turkey red at a material level, a more comprehensive examination is possible that allows for some persistent questions to finally be answered.

2

Global Exchanges and Anthraquinone Dyes

The precise origins of Turkey red are uncertain and undocumented, however, its names indicate potential provenance in Türkiye (*rouge turc, rouge d'Andrinople*) or the Indies (*rouge des Indes*), a now historical term encompassing a geographical area from India eastward to the island of New Guinea.[1] This vast region later gained the distinction "East Indies" after European explorers needed to distinguish them from the Antilles, the Caribbean archipelago which they assumed was the eastern side of the already-known Indies. These islands became known as the "West Indies." The name Turkey red predates the modern nation of Türkiye; both names derive from "the lands inhabited by Turkic people."

Records of historical practice and current textile traditions help trace the origins of Turkey red.[2] This chapter explores the geographical origins of methods fitting the definition Turkey red given in Chapter 1, how products and knowledge of practice were exchanged, and the beginning of production and industrialization in Western Europe. It also discusses madder and the other botanical dye sources that were used for centuries before the Chemical Revolution, when synthetic alizarin came to market and superseded them. The exchanges of material through trade, and of knowledge through intentional dissemination and industrial espionage, are a key part of the history of Turkey red and make a fascinating case study on the early days of industrial production, globalization, and industrial capitalism. The Turkey red industry in Western Europe during the nineteenth century is discussed in Chapter 5.

The Origin and Dissemination of Turkey Red

It is difficult to say where and when exactly Turkey red originated in the geographically vast and culturally diverse lands from which its names derive. Like art of dyeing itself, which was likely developed independently in multiple places at different times, there may not be a singular genesis for Turkey red.[3] Many significant developments in early human civilization arose in Mesopotamia and the Indus Valley, both of which were located between what is roughly present-day western Türkiye and northwestern India. One source says the complicated means to make Turkey red were known in India at the time of the Alexandrian conquests in the last centuries of the first millennium BCE.[4] Sea trade between Mesopotamia and the Indus Valley was established as early as the second millennium BCE.[5] Long-distance trade expanded during the early first millennium BCE with the emergence of iron metallurgy in the central Gangetic plains that span the northern region of the Indian subcontinent. Trade between the northwest, particularly Gujarat, and lands around the Indian Ocean from the Horn of Africa to Southeast Asia, was already established.[6] The analysis of dental calculus, the built-up food debris often discussed at dentist appointments, offers more evidence of extensive historical trade networks. Micro remains of individuals who lived in the Southern Levant in the second millennium BCE contained proteins from banana (*Musa*), indicating the genus was present there either through trade or established cultivation. *Musa* was domesticated in New Guinea in the fifth millennium BCE, and by the first millennium BCE had been dispersed under human cultivation as far as West Africa.[7] It is therefore entirely possible that textile materials and knowledge of their manufacture were also exchanged along these routes since the early days of long-distance seafaring trade. The map in Figure 2.1 illustrates the locations of places discussed in this chapter that are associated with Turkey red trade and production in the pre-Industrial era.

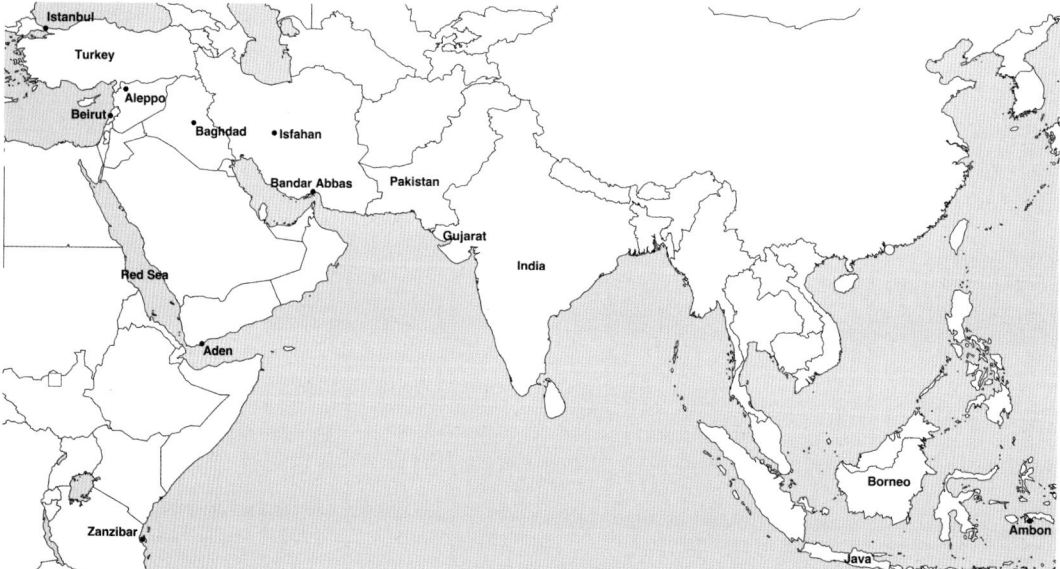

Figure 2.1 Map showing approximate locations of places associated with Turkey red dyeing and trade in the pre-industrial era. *Created with MapChart*

A 1767 French source says that Indian dyers were the first to dye beautiful and durable reds, which were later replicated by Turks, and this is how *rouge des Indes* became *rouge d'Andrinople*.[8] By the time it was dyed by Europeans in the mid-eighteenth century, the elaborate process was the result of countless experiments and adjustments by generations of dyers. Scholars in nineteenth-century Europe, who devoted extensive research to the composition of Turkey red and its origins, believed that the process originated in India. Debate ensued over whether *rouge des Indes* and *rouge turc* were the same based on material and technical differences.[9] The definition followed in this book is intended to overlook these minor distinctions in order to evaluate methods on a material level.

India

India is a geographically and culturally diverse country with a rich textile heritage too vast to be addressed in this book, though fortunately other texts provide more insight.[10] Ancient beads from a Neolithic burial dating to the sixth millennium BCE at Mehrgarh, Pakistan (located near modern-day Sibi) were analysed and found to contain mineralized cotton fibers, the earliest known example of cotton use in the Old World.[11] Although textiles are likely to decompose in burial environments, some conditions (like thread through metal beads) can preserve fibers through a process called mineralization where the organic material is replaced through chemical exchange by a more stable matrix, filling the same physical structure.[12] The impressions of cotton fibers can be identified due to their distinctive twisted ribbon shape (see Figure 2.2) even if they are materially no longer cotton. The evidence from the bead shows that makers on the subcontinent had millennia of experience working and experimenting with cotton.

Documentary evidence from the thirteenth century indicates Turkey red was being exported from Gujarat. The *Zhū Fān Zhì* (*A Description of Barbarian Nations* or *Records of Foreign People*), a thirteenth-century account by Zhào Rǔkuò (Chau Ju-kua following the Wade-Giles romanization), was written during the Song Dynasty (960–1279 CE).[13] The author was the Inspector of Foreign Trade in Fujian province on the southeast coast of China and lived from 1170 to 1231 CE.[14] Although he did not travel himself, he was well-positioned to create his account from his work and contact with those who passed through his port. Under the listing for Zanzibar (now part of Tanzania), Zhào says that every year traders from Gujarat and the Arab peninsula sent ships there with "white cotton cloth, porcelain, copper, and red cotton to trade."[15] Using records like trade accounts, ledgers, etc. as evidence of Turkey red requires some conjecture, and assumes that red cotton dyed before the late nineteenth century is probably Turkey red. In this case, the geography and the distance the goods were worth traveling for makes the red cotton likely to be Turkey red.

Gujarat is a modern state in northwest India that is primarily located on the Kathiawar peninsula (see Figure 2.3). Historically, the name has also served as a designation for the surrounding region, which has been part of many political entities over more than two millennia. Traders from Gujarat exchanged goods like metals, ivory, gold, pepper, gemstones, lacquers, and spices with people in Africa, Sumatra, Pegu (now Bago, Myanmar), Ceylon (present-day Sri Lanka), and the Moluccas (or Maluku Islands). Gujaratis dominated trade in the Indian Ocean from Aden, the port city in the south of modern-day Yemen, to Indonesia, sending vessels on an annual basis to faraway places.

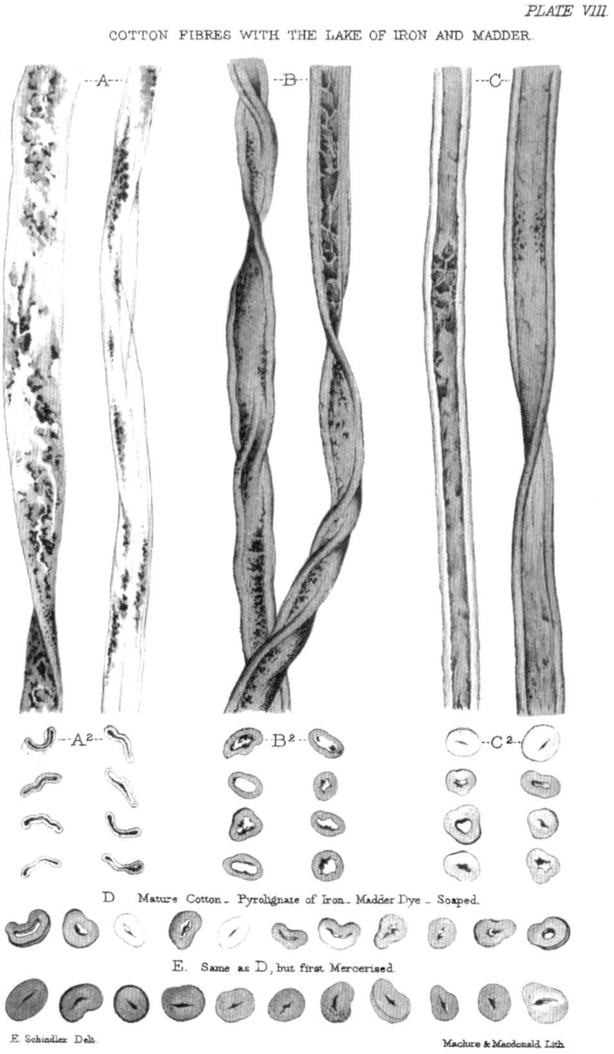

PLATE VIII.

COTTON FIBRES WITH THE LAKE OF IRON AND MADDER.

D   Mature Cotton - Pyrolignate of Iron - Madder Dye - Soaped.

E.   Same as D, but first Mercerised.

E. Schindler Delt.

Machure & Macdonald Lith.

Figure 2.2 Diagram of cotton fibers drawn lengthwise and in cross-section, showing the distinctive twisted ribbon profile. *XLIX. On the manner in which cotton unites with colouring matter. Crum, Walter. Journal of the Chemical Society, 1863 (16) 404–14*

Textiles were a major part of their economy and produced on a scale that enabled export.[16] Cotton was dyed red in Gujarat following a process that included an oil treatment (*kharāni*), tannins from myrobalans (*hardāvāum*), alum (*pārā karāvo*), dyeing (*rangavānū*), and finishing (*tapāvānū*). Dyeing was done with the local variety of madder, *Rubia cordifolia*, for which one of the local names in Gujarat was *majīth*. A type of purple called *jāmbolo* was made by including iron in the process.[17]

A range of local ingredients were employed, most notably the use of bark or plant material as an aluminium source. European and Middle Eastern dyers used the mineral source of aluminium

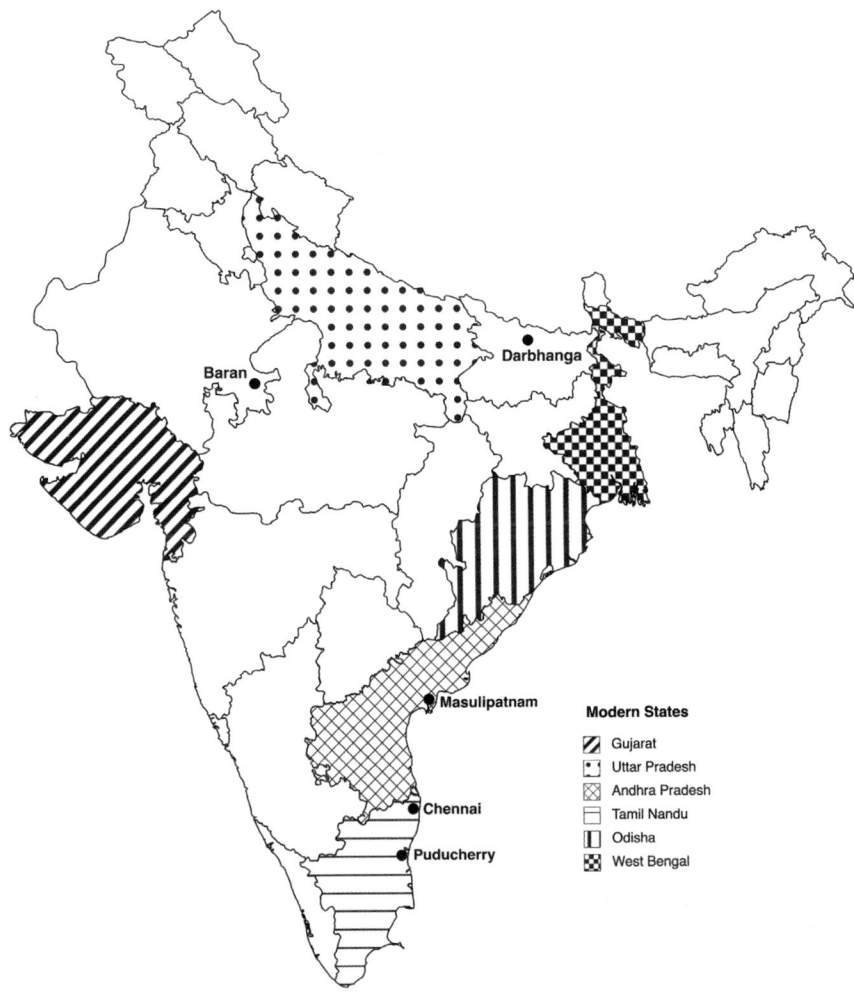

Figure 2.3 Map showing approximate locations of places in India associated with Turkey red dyeing. *Created with MapChart*

(alum, or potassium aluminium sulfate), while the ashes of aluminium-containing plants were often used in India and Indonesia. *Djirāk* was the name of a plant used for dyeing in India, which is probably the local name for a variety of *Symplocos*.[18] This genus accumulates aluminium in various parts of the plant tissue (species dependent), often at concentrations above one gram per kilogram of plant material.[19] Some dyers used *casha* leaves from the genus *Memecylon*, also known to accumulate aluminium, as a source. In the late nineteenth century, Félix Driessen, a Dutch Turkey red dyer and chemist, traveled to Java and studied local practice. He determined that *djirāk*, which he names as *Symplocos fasciculata*, was the aluminium source in red cotton dyeing. Driessen later presented his findings to the Société Industrielle de Mulhouse, a leading industrial and intellectual organization with many members in the textile industry.[20] Although published Turkey red methods use mineral alum, methods that use plant-sourced aluminium follow the same chemistry, which is how a material and process-based definition of Turkey red facilitates tracing its practice.

There is also evidence Turkey red was dyed on the Coromandel Coast, a historical European designation for the southeast coast of India where *kalamkari* originates. It was known for producing plain red cotton cloth, which came primarily from two regions. The first was centered around Masulipatnam toward the north, and the second further south around Pulicat and Fort St. George (now part of Chennai). Masulipatnam was a center of cultivation for chay root (*Oldenlandia umbellata*), and a lower quality was also grown in Pulicat.[21] Chay contains many of the same dyes as madder, discussed later in this chapter. A 1626 account by trader William Methwold says that red cotton yarn was traded from Masulipatnam to Tenasserim, in modern Myanmar, which was a shipping hub between the Bay of Bengal and Ayutthaya, in present-day Thailand.[22]

A 1742 account of dyeing in the Coromandel region by Père Coeurdoux, a Jesuit missionary in Puducherry (formerly Pondicherry), describes a Turkey red process. Half-bleached cloth was soaked in buffalo milk (rather than oil) mixed with powdered dried tannin-rich myrobalan fruit called *kadou* (*Terminalia chebula*), then dried in the sun. The cloth was treated in macerated goat or sheep dung, which helped lighten it and remove impurities, then washed. As with the tannins, the use of dung is optional for Turkey red. Their roles are discussed further in Chapter 3. The cotton was treated with mixture of alum and a particular kind of hard water obtained in Pondicherry, which supplied the calcium. Dyeing was done in tepid water with dried chay root. Finishing involved washes with dung and soap, then sun exposure.[23] Notably, this is evidence of cloth dyeing decades before it was achieved in Europe, further indicating its success is a matter of technique and experience.

Benjamin Heyne, a German-born botanist and surgeon who worked in Andhra Pradesh for the British East India Company, wrote an account entitled "Mode of Dyeing Red Cotton Yarn, Practised on the Coast of Coromandel" in 1795.[24] Heyne calls the method "exceedingly tedious and complicated" and says it was nearly identical to the one followed in the Levant. The process Heyne documents uses rancid "gingelie oil" (sesame oil), followed by a treatment of "casah leaves" (*Memecylon*) for the aluminium source. Dyeing was done with chay root, whose dried stalks were used as fuel to heat the bath. Drying in the sun played an essential part in the lengthy process, which took about a month to complete.[25]

Oil preparations for fibers are also documented north of the Coromandel Coast in Orissa (now Odisha) and West Bengal, although these regions are not as well known for producing Turkey red. It also appears at Darbhanga, in Bihar on the northeast border of India near Nepal, where a dark red color was dyed. A coarse type of cloth called *khārūā*, dyed with *āl* (*Morinda citrifolia*), was made in a few districts around Aligarh, Agra, Jhansi, and Jalaun, located in the present-day state of Uttar Pradesh in north-central India. *Morinda*, also known as *noni*, contains some of the same dyes as madder and chay. The preparation involved treatment with castor oil, sheep dung, tannins from myrobalan, alum, and then dyeing with *āl*. *Salu*, a fine red fabric used for turbans, curtains, and the borders of women's clothing, was dyed with *āl* following a similar process in the area around Jalaun.[26] One source says Baran, in north-central India near the border of Rajasthan and Madhya Pradesh, also dyed Turkey red.[27]

Indian textiles were traded with the Ottoman Empire and Western Europe via the Red Sea, and had a large market by the eighteenth century. Seafaring merchants facilitated the exchange of knowledge as well as goods, facilitating the growth of textile production in Ottoman cities like Istanbul, Cairo, Aleppo, Urfa, Gaziantep, and Diyarbakır, which began to produce their own imitations.[28] Competition increased with expanding markets and the dissemination of techniques. The textile economy in India changed significantly with the increase in direct trade with Europe and colonial expansion. Trade with India had been via routes managed by Venetians operating in the Persian Gulf from Bussorah (now Basra, Iraq), Baghdad, Aleppo, Alexandretta (now Iskenderun, Türkiye) and Beirut. In the early sixteenth century, Portuguese traders established their own by sailing around the Cape of Good Hope. The hub for Portuguese imported goods in Northern Europe was Bruges, then Antwerp (both in present-day Belgium), and after 1576, Amsterdam in the Netherlands. A 1580 edict by Philip II of Spain precluded trade with Lisbon, prompting the Dutch to establish their own routes to India. This led to the formation of the Dutch East India Company in 1602. English interest in textiles imported from India was piqued in 1587 when Sir Francis Drake captured a Portuguese trading vessel returning from India. In 1592, another Portuguese vessel captured by English privateers was returned to Dartmouth with a cargo containing calicoes, lawns, quilts, carpets, and other commodities, motivating the English to establish direct trade with India via the London East India Company in 1599. The French East India Company was granted a charter in 1664.[29]

European trade in imported textiles was a precursor to European industry replicating those goods—if trade were so profitable, what could be earned by controlling the means of production? This effort was accelerated by the Industrial Revolution, which began in the mid-eighteenth century. By the early nineteenth century, mechanical innovations had significantly raised production capacity over that of hand production. The implementation of water- and coal-power was the final step toward industrial production, with which artisans could never compete.

Indonesia

Rouge des Indes could also imply Indonesian origins for Turkey red, though it is more likely the technique had disseminated eastward from India since there is less evidence for a strong tradition of dyeing red on cotton in Indonesia. Javanese chronicles say that around 603 CE a ruler of Gujarat, warned of the approaching destruction of his kingdom, dispatched his son to Java along with five thousand followers in many vessels, where they founded a great civilization. An inscription at the Plaosan temple in Central Java refers to a "constant flow of people from Gujaradesha," for whom the temple had been built around 800 CE. Gujaratis also settled around Gresik, on the north Javanese coast.[30] This diaspora of Gujaratis to Java could have brought with them knowledge of dyeing techniques and established similar practices in their new home.

Although it may not have been made on a large scale, a late-seventeenth century account by Georg Rumphius, a German-born botanist working for the Dutch East India Company in the Ambon archipelago in eastern Indonesia, describes red yarn dyeing. Rumphius identified local plants used as aluminium sources, and described dyeing cotton and linen red with an oil pre-treatment. The color came from *mengkoudu* (*Morinda citrifolia*). A 1707 manuscript in the collection of the Koninklijk Instituut voor Taal-, Land- en Volkenkunde (Royal Netherlands Institute of Southeast Asian and Caribbean Studies) describes a dyeing process with steps consistent with a Turkey red process using plants with local names. Batik, the wax-resist dyeing technique closely associated with Indonesia, could be dyed on a Turkey red ground by applying darker dyes (blacks, browns) over it, and was practiced in Semarang, on Java from the eighteenth century. Evidence indicates the batik was applied to cloth imported from the Coromandel Coast, showing dyeing was better established there.[31]

The island of Borneo lies north of Java and is divided today between the countries of Indonesia, Malaysia, and Brunei. It is home to the Dayak people, a name that encompasses many cultural subgroups. A nineteenth-century Dutch account describes some Dayak textiles of notable fastness that were made by steeping the fibers in an oil or fat, followed by a treatment with plant juices, and that the color on oiled fibers was better than on unprepared ones. The plant used for dyeing is not named, but the final color was a deep brownish yellow rather than red. Although they were not dyeing red, the practice shows knowledge of preparatory oiling treatments and their application.[32] Dusun Dayak weavers in Kalimantan still perform the *ngaos* ceremony, which includes a complex oiling process involving rancid coconut, oil seeds from many tree species, and sometimes a small amount of fat from a ritually significant animal source like monitor lizards, freshwater turtles, chickens, fish, snakes, or crocodiles. Dyeing was done with *Morinda citrifolia* or another species to create red.[33] It is not clear when the Dayak began dyeing Turkey red, but it is not surprising the practice spread throughout these islands already engaged in seafaring trade. A nineteenth-century Dayak ikat, shown in Figure 2.4, contains red cotton yarn which may have been dyed in this fashion, although it is also possible from the date that the yarn was imported from Europe and woven locally. The variegated red is intriguing, and merits further technical investigation. In Figure 2.5, a colorized photograph from the early twentieth-century publication *Peoples of All Nations* shows two young Dayak in formal garments. The boy is wearing a red cloth with a bright yellow pattern characteristic of Turkey red made in Europe and exported for the Pacific market (see Chapter 4).

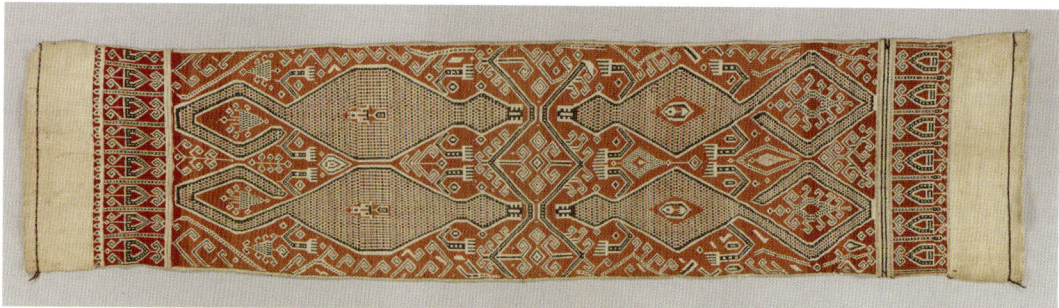

Figure 2.4 An Iban ceremonial cloth from the nineteenth century woven in cotton. The red yarn was likely dyed following a Turkey red process, which was practiced in Indonesia, however by this time it is unclear whether it would have been made locally or imported for use by the weaver. *Metropolitan Museum of Art. Gift of Dr. Joel Confino and Lisa Alter. 2016. 2016.736.3*

Figure 2.5 An early twentieth-century colorized image of young Iban, or Dayaks, in gala dress. The red and yellow print worn by the boy looks very much like an imported Turkey red. *Peoples of All Nations. Charles Hose, 1922*

The Levant and the Ottoman Empire

As discussed earlier in this chapter, trade between Europe and India flowed through the Eastern Mediterranean before direct sea routes were established starting in the late sixteenth century. This included the exchange of Turkey red goods and knowledge of its process. The climate there was also suitable for growing madder, and olives for the oil treatment. The south and west of Türkiye, northern Syria, and Cyprus were the primary places of local cotton production from the fourteenth century. Egypt, Syria, Türkiye, and the Balkans also cultivated cotton for a long time, appearing in Ottoman records by the sixteenth century. Cotton became valuable for making sailcloth, garment linings, and undergarments, mostly for the military. By the seventeenth century a range of cotton products was available, expanding to include civilian consumers. Finer products were used for sewing and weaving borders on high-quality garments, while coarser ones were for candle wicks and more utilitarian objects. Turkish centers of cotton production were Şanlıurfa and Diyarbakır, in the south of the country. The availability of essential materials, and some knowledge of cotton processing, facilitated the establishment of Turkey red dyeing.

There is no evidence of exactly when Turkey red dyeing was established west of India, and the long history of trade leaves a broad window of possibility. The region, which was largely part of the Ottoman Empire from about 1299 to 1922 CE, was home to a diversity of people, cultures, and economies that are often treated collectively in analysis. It may have migrated following the conquests of Alexander the Great (336 to 323 BCE), disseminating slowly throughout Macedonia and Thessaly in present-day Greece. Trade caravans and pilgrimage routes are other potential avenues of knowledge and material exchange between disparate populations. Archaeological finds at Qasr Ibrim, located in the south of Egypt near present-day Aswan, include a child's cap containing red cotton and a narrow red cotton belt. The site was continuously occupied for at least 3,000 years and abandoned in 1811 CE. Estimated dates for the objects are not provided, so there is a small chance the yarn is a European import if the objects are from the last half-century of its activity. An archive document from 1519 CE says that Turkish dyers first brought the process to Bursa in the thirteenth or fourteenth century from their home in Central Asia. Although a minority opinion, it has been proposed that Turkey red originated in Central Asia and spread to India. Levantine dyers obtained the best quality alum, the mineral source of aluminium, from Tashkent, the capital of modern Uzbekistan. Central Asia is home to many shepherding cultures, and early tanning methods for hides often involved treatments with fat and alum. Leather goods were also made in India, however, so this knowledge was not exclusive to Central Asia. The translation of tanning practice to cotton dyeing is more easily accomplished by settled peoples than nomadic shepherds since dyeing is not a portable craft. It involves immersing or saturating the fibers in an aqueous bath containing the dissolved or dispersed coloring material, consuming large volumes of water often with heat applied. These requirements are ill-suited for a nomadic lifestyle, which is not likely to facilitate the development of such a complicated technique as Turkey red. While possible that Turkey red originated from Central Asia, Tashkent was situated on a trade route to India and it is more likely that while dyers may have come from Central Asia, the dyeing technique came from India via their home.

The expansion of the Ottoman Empire disseminated goods, knowledge, and skills. Bursa, in the northwest of present-day Türkiye, functioned as a hub and was the destination of many caravans from the east (see Figure 2.6). Indian cottons became popular among wealthier Ottoman customers in the seventeenth and eighteenth centuries. Ottoman fabrics were of a lower quality than Indian ones, and local production could not satisfy demand. This caused a trade imbalance with more coin leaving the Ottoman Empire for India, which of course piqued interest in improving local production. It was eventually dyed in İzmir and Edirne, located near the border of Türkiye and Greece and formerly called Adrianople (hence *rouge Andrinople*), and in Aleppo, in particular red *boğası*, a high-quality cotton twill fabric. *Boğası* was a popular export to Marseille, where in French it was called *boucassin*. In addition to their trade activity, Armenians also dyed Turkey red in Astrakhan, on the Volga Delta near the Caspian Sea, using fish oil. This was done outdoors and

Figure 2.6 Map showing approximate locations of places in Western Asia and Europe associated with Turkey red dyeing. *Created with MapChart*

seasonally, from the spring to late autumn. The embroidered border in Figure 2.7 was made in Russia in the late eighteenth century using red cotton yarn that was probably dyed around the Black Sea coast.

From the seventeenth century, Ottoman cotton fabrics were exported to the north and west of the empire around the Black Sea and into Eastern Europe along the Danube, traveling as far as Poland. From Smyrna (now İzmir), products went via the Aegean Sea toward Italian ports like Venice, a distribution center for Central Europe, and overland via Macedonia. Much of this trade, which formed part of a long caravan route that began in India, traveled through Iran and the Ottoman Empire, and finished in Marseille, was facilitated by Armenian merchants who also dispersed knowledge of manufacturing techniques. In the early seventeenth century, Armenian immigrants had a large settlement in New Julfa, which is now the Armenian quarter of Isfahan in Iran. As Eastern Christians, they were well-positioned as to facilitate exchanges between the East and West, and they played an active role in trade across Asia and into Europe. One such route extended from Isfahan to Bandar Abas on the Persian Gulf Coast via Shiraz, then across the water to Surat in Gujarat, Aurangabad, Brahmapur (Berhampur), Sironj, and Agra in north-central India. With all these connections, it is no surprise knowledge of Turkey red dyeing spread westward.

Documentary evidence of red cotton trade shows the geographical extent of its use. Caravans from Cairo traded red cotton yarn and bales of red cotton fabric to Sennar (present-day North Sudan). It may have been dyed in Cairo, or traded from further afield. Muslim pilgrims from Tunis, Algiers, and Tripoli in North Africa undertaking the *Hajj* traveled to Mecca and returned with red cotton cloth and thread. Trade into Europe probably traveled from Isfahan via Trabzon (historically Trebizond) to Kaffa (now Feodosia in Crimea) or Mangalia (southeast Romania). From Mangalia the route went north to

Figure 2.7 A decorative border made in Russia, late eighteenth century, with embroidery and trim in red cotton against a plain linen ground. The red cotton is likely Turkey red based on the color and surface wear, and perhaps came from Astrakhan on the Caspian Sea coast. *Brooklyn Museum Costume Collection at The Metropolitan Museum of Art, Gift of the Brooklyn Museum, 2009; Gift of Mrs. Edward S. Harkness in memory of her mother, Elizabeth Greenman Stillman, 1931. 2009.300.3426*

Iaşi, then Lublin, in Poland, and finally Warsaw. This exchange of goods coincided with the development of capitalism and a bourgeois class, and imported textiles were a major part of this new economy.

Turkey red dyeing slowly spread westward into Central Europe, mostly through the migration of experienced practitioners. The westward expansion of the Ottoman Empire into Greece brought Turkey red and later, Greek dyers played a key role transferring the process across Europe. It is difficult to determine exactly when dyeing first occurred, but Ottoman activity in Greece began in the fifteenth century. In 1423, Thessaly was conquered by the Ottoman general Turahan Bey, a gifted military commander adept at establishing stable, peaceful administrations following his conquests. Bey implemented madder dyeing in Tyrnavos. Although it is not confirmed whether this was initially Turkey red, it would at least have been the foundation for later practice.

The town of Ampelakia (also Ambelakia) in the region of Larissa, southwest of Thessaloniki, was once a center of Turkey red dyeing. From there, yarn was traded to Vienna, Buda, and German territories to the north. The industry was organized by Georgios Mavros, whose surname means "black" in Greek. He engaged with Viennese trade to the point that he was often known as Schwartz, which is "black" in German. The Schwartz mansion, built with their Turkey red fortune, still stands in Ampelakia as a historical site. The industry in Ambelakia dominated the trade in dyed cotton yarn to Central Europe during the second half of the eighteenth century, peaking in the late 1780s before French manufacturers eroded their market dominance. Turkey red was also dyed in the nearby towns of Agia and Tyrnavos. In the late eighteenth century, French traveler and Consul General in Greece Baron Félix de Beaujour observed many towns and villages in the region around Mount Ossa and Pelion engaged in Turkey red dyeing using olive oil.

Imported wares began to threaten Ottoman production by the late eighteenth century, and some effort was made to adapt practice to respond to the growing competition. Initial trade contact between European merchants and the Ottoman Empire was to find goods to purchase, rather than a market to sell European products. This flow began to reverse as, as it had somewhat between the Ottomans and India. European manufacturers found ways to imitate the textiles they had formerly purchased by cheaper, mechanized means, which they could then sell back to the very markets for which they were once customers. Furthermore, the rapid increase in raw cotton exports to Europe deprived the local industry. Ottoman textile manufacturing went into decline in the early nineteenth century as hand production was unable to compete. The diffusion of technical knowledge aided European advances, while Ottoman regionalized state authority constrained labor and capital. Toward the end of the eighteenth century, the government attempted to intervene by prohibiting the import and use of Indian cotton, but the dual threat of Indian goods and the growing European export market was too much. Some adaptations occurred, like importing machine-spun British yarn to dye for sale to Russia, but hand dyeing was no match for machinery. With this transition, the global Turkey red market became dominated by Western European products.[34]

The Hapsburg Empire

The western border of the Ottoman Empire was, for some time, the eastern boundary the Hapsburg Empire, which ruled various parts of Central Europe from the thirteenth to twentieth centuries.

Initially, Turkey red was imported from Ottoman dyers, including those in Ampelakia. A 1784 estimate from the earliest book in German about Turkey red says that two million pounds of yarn (approximately 900,000 kg) were imported annually to Germany, Hungary, and Bohemia via Trieste, in turn diverting two million thalers back to the Ottomans.[35]

As always, interest developed in making it locally to raise profitability. Attempts to dye Turkey red were made in Transylvania and the Banat of Temesvar, hereditary provinces of the Hapsburg Empire located in present-day central and western Romania. Merchants in Temesvar (now Timișoara) invited a dyer named Adam Karathanasis to establish a factory there, which operated for a few decades in the mid-eighteenth century. Hapsburg rulers tried to foster the growth of local cotton industries, including red dyeing, in the following decades as demand rose for cotton goods. In 1755 a "factory association" of Greek dyers was granted financial support, privileges, and customs concessions with exclusive rights to dye Turkey red yarn for twelve years. They also received preferential terms for raw material imports within the Hapsburg territory if they took young Germans as apprentices and created a local repository of knowledge. They established a workshop in Vienna near Schönbrunn palace with a theoretical capacity to dye five thousand pounds of yarn daily. In 1757, an Ampelakia dyer named Panayiotis Vengelinos applied to Hapsburg authorities for privileges to run a Turkey red workshop. He submitted his method as part of the application, with the understanding that it would be kept secret. Vengelinos was moderately successful, but had difficulties with raw material supply and labor costs. Once his privileges expired around 1766, he relocated to Saxony and unsuccessfully attempted to start another dyeworks.

The late 1760s to mid-1770s were an active period for craftsmen and Austrian officials trading in the secrets of Turkey red dyeing and establishing operations, taking advantage of the demand for red cotton yarn. Dimitrios Schwartz, of the Ampelakia family of dyers, operated a red dyeworks in the Leopoldstadt quarter of Vienna from 1784 to 1788. It closed either due to bankruptcy, or because it was easier to continue importing from Ampelakia. Ioannis Kyriazopoulos operated a Turkey red works in Sternberg (now Šternberk, Czechia) and Baden bei Wien in Austria around 1812. Operations owned by Greek dyers Ioannis Nikolaou and the Vlachos brothers flourished in Trieste, near the present-day border of Italy and Slovenia. Their yarn was exported to Lombardy and southern Germany, farther than product from Ampelakia traveled. The Vlachos brothers had learned dyeing from their father in Thessaly, and had lived in Marseille and Montpellier before settling in Trieste.[36] Despite this activity and the government support, there is little indication that Turkey red works established by Greek dyers in the Hapsburg lands ever achieved the success or scale of production that the French and British industries would in the latter half of the eighteenth century.

France

French efforts to learn Turkey red dyeing began in the early eighteenth century. Like textile manufacturers in the Hapsburg territory, their counterparts in Rouen purchased Turkey red yarn from Ottoman suppliers.[37] Naturally, they sought to profit by dyeing it domestically. Toward the middle of the 1700s, Louis Sauveur, the Marquis de Villeneuve and ambassador to the Ottoman

Empire, sent recipes to France.[38] The migration of Greek dyers was the major factor in the dissemination of Turkey red, however, and finally enabled the establishment of a French industry.[39] On August 26, 1747, exclusive rights to dye Turkey red in the kingdom for twenty years were granted to Rouen entrepreneur Pierre d'Haristoy and dyer André Fesquet, and a manufacturer in Languedoc named François Goudard. They established a dyeworks at Darnétal near Rouen, and at Aubenas in the south (see Figure 2.8). The founders engaged Greek dyers from Smyrna via Marseille, who brought their practical knowledge of Turkey red to France. The three partners were supported in their endeavor by investment from Rouennais businessmen and Jean Hellot, a chemist at the Académie des Sciences. They had all gained knowledge of Turkey red dyeing independently of each other and agreed to cooperate on the initiative at the behest of the French government, who offered a subsidy in exchange for exclusivity and secrecy. To free the local industry of dependence on imported raw materials, the director of the botanical gardens in Rouen attempted to cultivate madder. The state made concessions to facilitate the emigration of artisans to work in French

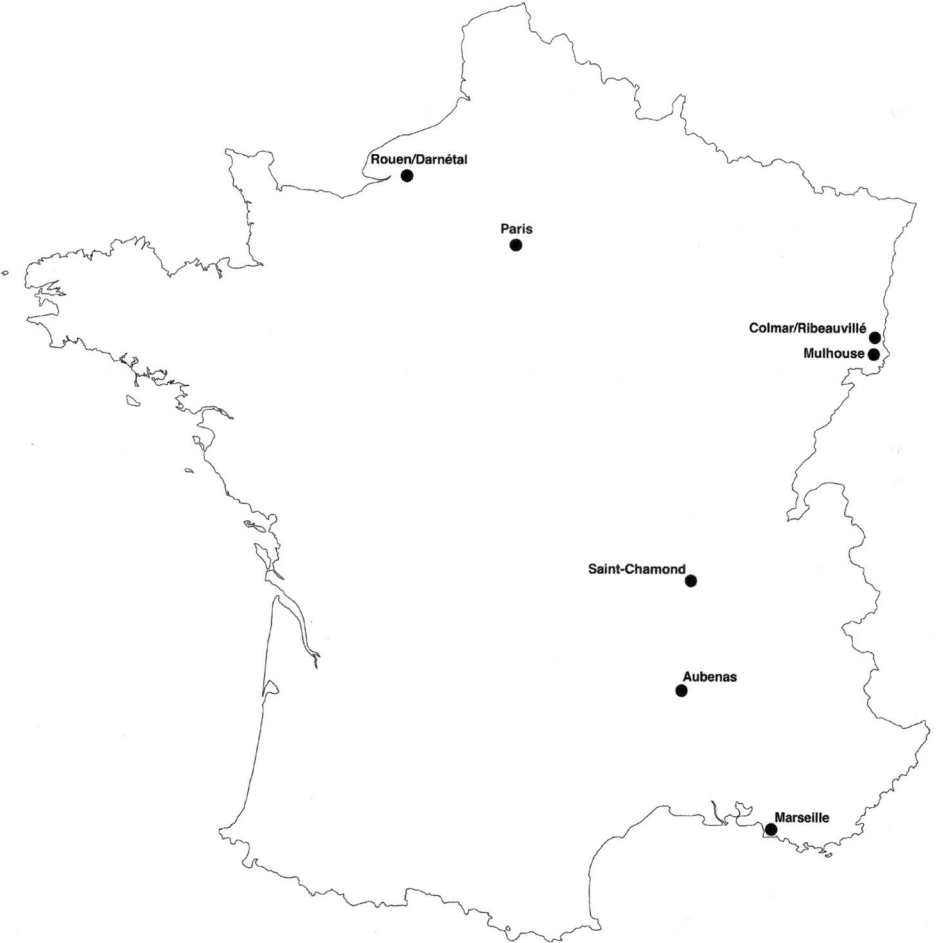

Figure 2.8 Map showing approximate locations of places in France associated with Turkey red dyeing. *Created with MapChart*

Turkey red works, providing them with legal status and residence that mitigated potential issues with local guilds. Unfortunately, they could not produce enough acceptable product to meet demand. The initial venture in Darnétal was unsuccessful due to trouble selling product and tension between the partners, and did not survive its twenty years of exclusive rights. One source says John Holker, a Jacobite exile from Lancashire, opened the first Turkey red dyehouse in Rouen, though there is more evidence it was d'Haristoy and partners. Later, two Parisian merchants named Pouce and Archalat also obtained government support to hire Greek dyers from Edirne, and established a workshop in Darnétal.

Around the same time as d'Haristoy, Fesquet, and Goudard were opening their dye houses, a merchant named Jean-Claude Flachat from Saint-Chamond, southwest of Lyon, received authorization to go to Constantinople (now Istanbul) to study the Turkey red industry and learn their methods. He returned to France in 1756 with documentation of the process. In 1765, the French government published the first widely-disseminated method for dyeing Turkey red, contrary to the wishes of manufacturers who had managed to establish operations since the original Darnétal venture failed. In 1766, Flachat published his two-volume book *Observations sur le commerce et sur les arts d'une partie de l'Europe, de l'Asie, de l'Afrique, et même des Indes Orientales* (Observations on the commerce and arts of part of Europe, Asia, Africa, and even the East Indies). Among many other things, he discussed the state of Turkey red dyeing in France to date and described the process, a verbatim reproduction of the 1765 government publication. Since his journey to investigate Turkey red dyeing was government-supported, it is likely that he is the unnamed author of the 1765 text and chose to include it in his own book the following year. Flachat received permission to establish a *manufacture royale* in Saint-Chamond, a designation granting royal privilege and exemption from certain statutes, which was tasked with training as many students as possible. He also brought experienced Greek dyers to the firm, who were entitled to become naturalized French citizens after three years. In the following decades, more Turkey red works opened in the south of France. The first in Marseille, located in Bachas close to the Bassin d'Arenc, was established in 1770 by Fistler frères. By 1780, there were Turkey red dyers in nearby Aix, Aubagne, and Lambesc. In Aix-en-Provence, there is a street at the southern end of the city named *Chemin du Coton Rouge*. A Turkey red works was established in Nîmes, northwest of Marseille, by an Eymard. In 1776, the Conseil d'Etat authorized a manufacturer named Quinquet to dye Turkey red in Montmartre, Paris.[40]

Britain

Cotton textile manufacturing in Britain grew slowly at first due to import and production bans on certain cotton goods in an effort to protect local wool and silk industries which were threatened by the increasing popularity of "painted," or printed cottons (often called *indiennes*) first imported from India and then France. They became popular from the mid-seventeenth century for their lighter weight and lively prints relative to heavier, expensive patterned silks and wools.[41] In 1700 an Act of Parliament was passed prohibiting Indian silk and printed cotton fabrics for domestic use in apparel and furniture at a penalty of £200 to the weaver or seller. Demand for printed cottons generated an underground economy, prompting further regulations to be passed.[42] In 1720 a prohibition against the domestic use of cotton printed in England went into effect (it was still made for export, and still allowed in Scotland).[43] Fustians (cotton-linen weave) were permitted, but linen fibers limited the range of colors that could be achieved since it did not take the available colors as well as cotton. The Act of 1720 was repealed in 1774, allowing English calico printers to sell on the home market, though duties were applied until 1832.[44] Restrictions limiting manufacturing locally inhibited the development of local knowledge on cotton dyeing, delaying the possibility of successfully imitating a specialist product like Turkey red.

Nevertheless, following the established pattern of import to imitation, British manufacturers eventually sought to make their own Turkey red. In 1732, weavers in Glasgow and the Gorbals, a village south of Glasgow Green later annexed by the expanding city, founded the Red Society. Their minutes book in the Mitchell Library collection states the purpose of their organization was "for the better regulating of the dying of red linnen & cotton yearn madder red," and some members were appointed to "buy goods and sufficient deye stuff of any kind in bulk as the said Company shall allow."[45] The Charter of the Red Society was drafted in 1759, outlining a charitable institution whose aims included "Encourageing the Dyeing of Mader Red."[46] In 1738, a middleman in the textile trade wrote to London clients about the difficulty of dyeing a good red on cotton.[47] The Secretary of State received a petition in 1751 for a grant or transport to Manchester from a Dane claiming knowledge of Turkey red dyeing, though nothing more appears to have happened after this. In 1756, the Society of Arts offered an award for the successful dyeing of Turkey red, with some applicants but little success.[48] John Wilson, a velvet and cloth dyer in Ainsworth, Lancashire, was probably the first person in Britain with practical knowledge of Turkey red, which he deemed tedious. Wilson learned the method by sending an agent to Smyrna in 1753 at his own expense, an early act of industrial espionage. The young man stayed with a merchant named Richard Dobs, who spoke Greek and facilitated his entry into the local dyeworks. In his 1786 account, Wilson describes why Turkey red was not suited for his business and comments "this valuable Colour cost me several Hundred Pounds."[49] Nevertheless, he received a prize of £50 in 1761 (about £5,000 today) from the Society of Arts for dyeing the best Turkey red, though the fastness of Wilson's product still did not compare to imported yarn.[50] The Society also paid £100 to a Mr. Simon Spurret of Isleworth for discovering a method of dyeing cotton yarn Turkey red, though his practice seems to have been adaptations on madder dyeing rather than a true Turkey red process.[51] An article in the *Manchester Mercury* on March 17, 1767 says a dyer from Glasgow was purportedly experimenting in Manchester with unknown results. A 1769 letter by the famed Scottish inventor James Watt references "a curious book ... unluckily in German" that a Swiss dyer, who had been successful in dyeing "standing red on linen or cotton," was helping him read.[52] No other evidence of these events is known. Despite interest from dyers and trade organizations, none of this led to the establishment of a Turkey red industry in Britain.

A minor debate ensued in the nineteenth century about whether the first dyer of Turkey red in England was a man named Malachi Hawtayne, who was outfitting the English army in Wandsworth, Surrey. He obtained knowledge of his process, described as "scarlet, or turkey-red," from his wife, whose family were experienced Dutch textile workers that had immigrated to the area.[53] The term scarlet dyeing often referred to dyeing with cochineal, and military garments were likely to be woolen rather than cotton, so he was likely dyeing a Turkey red color rather than cotton.

Madder

The history of Turkey red is closely linked with that of its original dye source, madder root. It is one of the oldest and most commonly used dyes in Europe, the Middle East, and India.[54] Of all the natural dyes available, there was "none so important or so valuable as madder," which dyed colors fast to washing, light, and rubbing.[55] Dyer's madder (*Rubia tinctorum* L.) is native to Central Asia and the Caucasus. Cultivated since antiquity, it is mentioned by historians Dioscorides and Pliny the Elder, who describe its use by Egyptians, Persians, and Indians. Ancient Greeks and Romans knew it as *Erythrodanon* and *Rubia*. The Crusades, which began in the late eleventh century CE, brought madder cultivation into Western Europe. Names for madder in other languages include *garance* (French), *meekrapp* (Dutch), *krapp* (German), and *rubia* (Italian).[56] It has been identified analytically on archaeological textiles from the Egyptian site Tel-el-Amarna, dated to around 1350 BCE, and in textiles from Mohenjo-Daro as early as 3000 BCE. The earliest written directions known for dyeing with madder are found in the Papyrus Graecus Holmiensis, from about 300 CE.[57] The history of madder as a colorant and the economy of its trade is a story of its own, much greater than its part in Turkey red. A more comprehensive discussion can be found in Chenciner's *Madder red: A history of luxury and trade*.[58] Figure 2.9 shows a 1739 botanical illustration of madder by the Scottish illustrator Elizabeth Blackwell.

The Cultivation of Madder

Multiple species from the genus *Rubia* function as dye sources. *Rubia* are native to a number of biomes around the world, and grow in Mexico, Europe, Siberia, China, Japan, India, Persia, and Asia Minor. In addition to dyer's madder *Rubia tinctorum* L. (also called *R. sativa*, *R. sylvestris*), wild madder (*Rubia peregrina* L.), Indian madder (*Rubia cordifolia* L. or *R. munjista*), and Japanese madder (*Rubia akane*) were used.[59] *Rubia tinctorum* was the most desirable for dyeing, followed by *Rubia peregrina* and to a lesser extent *Rubia cordifolia*. As discussed earlier in this chapter, Turkey red was also made with other red dyes like chay and morinda, though these do not feature as significantly as madder in historical texts. Figure 2.10 shows dried, cut roots from *Rubia tinctorum* L.

Madder was grown in Rome and near Ravenna, in the north of Italy near the Adriatic Sea, from the first century CE. In the seventh century it was grown near Paris, and is mentioned in the estate records of Charlemagne.[60] It was established in the south of France, Alsace, the Netherlands, Silesia, Saxony, Bavaria, Baden, the Lower Palatinate, Belgium, Austria-Hungary, Tuscany, Sicily, Spain, Greece, Romania, and Russia. Britain never grew it on any significant scale since the climate is too damp and heavy rains damage the roots, though there were attempts to do so as early as the seventeenth century.[61]

Wars during the sixteenth and seventeenth centuries suppressed madder cultivation in France, leaving the Netherlands with an effective monopoly for some time. It was re-introduced to Alsace and the Comtat Venaissin in Provence, both in part due to government support and the work of an Armenian immigrant named Jean Althen, who came to France from New Julfa in 1756 and recognized the favorable agricultural conditions for madder. In the Netherlands, madder was

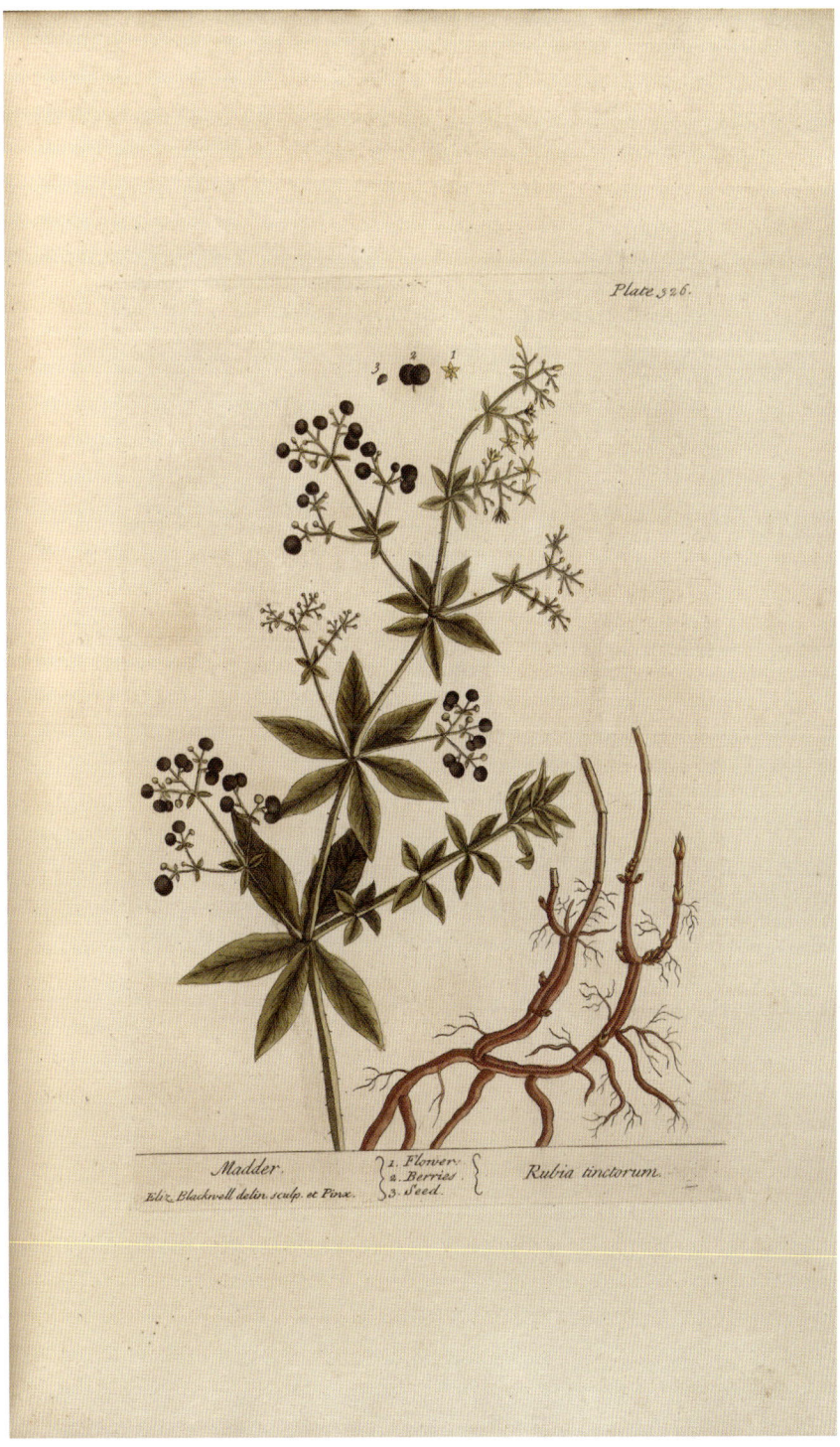

Figure 2.9 Botanical illustration of madder plant engraved by Elizabeth Blackwell, 1739. *George Arents Collection, The New York Public Library. "Madder" The New York Public Library Digital Collections. 1739*

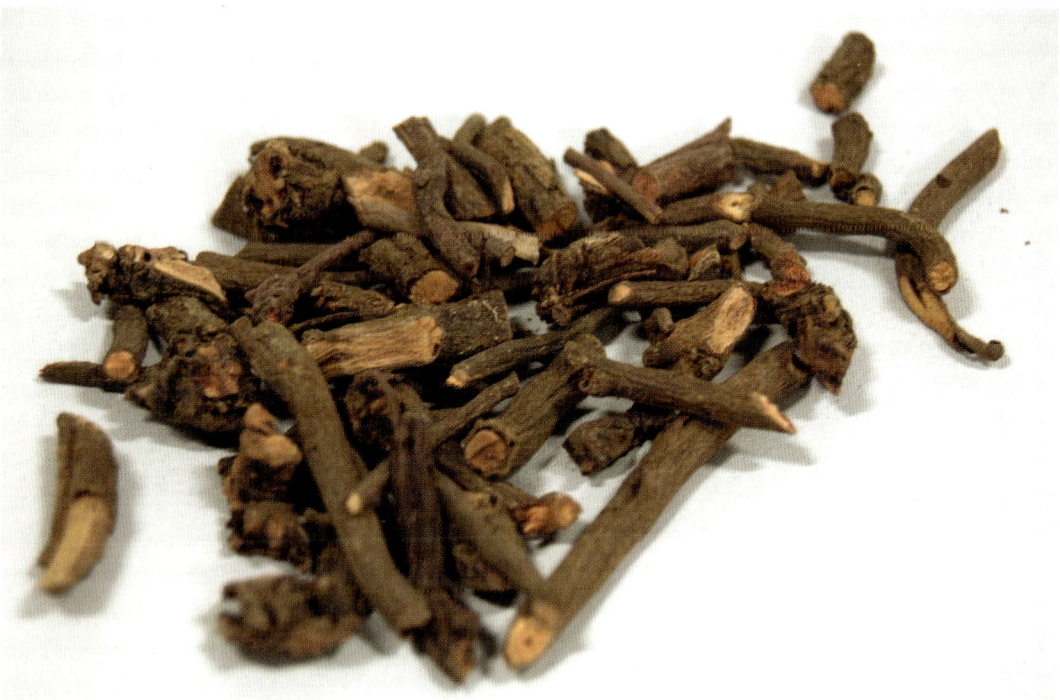

Figure 2.10 Dried and cut madder roots show the reddish tone of the woody interior where the anthraquinone dyes are concentrated. The roots, which would be ground into a powder for maximum extraction, have an earthy, tea-like odor. *Julie Wertz*

grown primarily on polders, or land reclaimed from the sea, in southwestern province of Zeeland. Cheaper, lower-quality madder was grown in Saxony and Silesia, around the present-day border between Germany and Poland.[62]

To supply the British textile industry, a considerable amount of madder had to be imported—Turkey red methods typically specify one to two pounds of ground madder per pound of cotton to be dyed. France and Netherlands were the major suppliers. A 1758 treatise on the cultivation of madder by English botanist Philip Miller discusses its importance in textile production, the dependence on Dutch trade, and concerns with rising costs and potentially adulterated product. He describes practice in the Netherlands, which grew very good madder, and includes illustrations of equipment and buildings. Miller acknowledges madder production in England is low and that some doubt its viability, concluding that failed trials to cultivate madder are the result of a lack of skill.[63] At the time of Miller's treatise, the annual import value of madder from the Netherlands to England was valued at £30,000 (about £3 million in twenty-first-century currency).[64] The text was available in Philadelphia by 1762, but madder was never grown on a commercial scale in the United States. There are only two known accounts of what could be considered commercial ventures, but they were short-lived and no greater than thirty acres in size.[65]

Although madder was grown in many places, roots from Smyrna and Cyprus were the best. They were about 10–25 cm long and thicker in diameter than European madder, partly because they were grown for four to five years instead of two to three like the second-best ones, from the Netherlands.[66] Good roots range in diameter from the size of a writing quill to a little finger and should have a reddish color, strong odor, smooth bark, and be of at least two to three years' growth.[67] In the Levant, madder growers trained the plants like beanstalks to increase stem growth, which also increased the root size. Dutch farmers kept the roots small to reduce flood risk from destabilizing the soil on the shallow polder land.[68]

Madder roots were usually dried after harvest. Some dyers said that fresh madder was better, but the volume was inconvenient to transport due to the weight of water present, which also made it prone to spoilage. The major consensus, however, was that its use as a dye improved after ageing. In Northern Europe, madder roots were dried in a furnace that was vented periodically to release humidity. The roots were spread on multiple levels, which were rotated downward toward the heat source as they finished drying. Madder was sun-dried in the Eastern Mediterranean, which improved its dyeing power since it was done for longer at much lower temperatures. Dry roots were repeatedly flailed, sieved, and winnowed to remove bark and woody material. The finest dust was discarded, and the madder fragments sorted by sieves with the largest fragments reserved as the best. These were heated again until they could be snapped, then milled and sieved to remove bark and sieved again before milling into a powder. In the Netherlands, madder roots were processed in a similar fashion, and then the dry powder was packed tightly and sealed in casks for storage in Rotterdam warehouses to develop the color. Stored ground madder should be protected from humidity, since absorbed moisture will degrade it. Adulterants like dirt and brick dust were frequently mixed with ground madder, so in the Levant it was customarily sold unmilled to demonstrate authenticity.[69]

Madder Composition and Derivatives

Anthraquinone (see Figure 2.11) is the basis for a major class of dyes, of which madder components alizarin and purpurin are members. Figure 2.12 shows alizarin (1,2-dihydroxyanthraquinone) and purpurin (1,2,4-trihydroxyantharquinone). Research on the identity and structure of madder anthraquinones began in the nineteenth century, and continues today.[70] Many natural anthraquinones are known, although not all function as dyes. Cochineal and kermes are other significant natural sources of anthraquinone red dyes, with distinctive components not found in madder. The main colorants in madder are alizarin, purpurin, pseudopurpurin, munjistin, anthragallol, xanthopurpurin, lucidin, and rubiadin.[71] Madder is a polygenetic dye, a property that means it can produce several colors depending on the mordant used.[72] The color of madder dyeing varies greatly depending on the presence and relative amounts of the various colorants.[73] Alizarin combines with aluminium, iron, chromium, and tin, or a combination of these, to produce colors including red, rose, black, violet, lilac, and dark brown.[74] A mid-nineteenth-century cotton print shown in Figure 2.13 was mordanted with iron and aluminium, then dyed with madder, demonstrating the range of colors it can produce.

Due to its economic value, there was a keen interest in madder cultivation, composition, and how to improve its use. Research in the nineteenth century facilitated a shift from "craft-based to science-based industry," as processes became better understood and more quantifiable.[75]

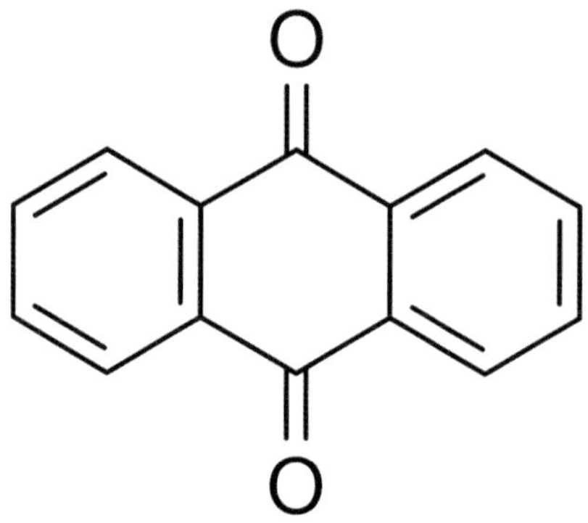

Figure 2.11 Structure of anthraquinone (9,10-anthracene).

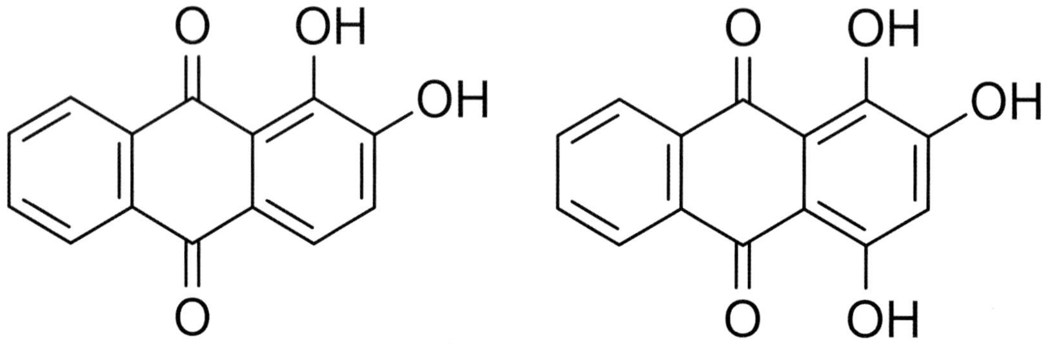

Figure 2.12 Structures of alizarin (left) and purpurin (right).

They found calcium in the soil improved the quality of madder grown, such as that of the Vaucluse district in southeastern France, which was light and calcium-rich.[76] Practical experiments included transplanting Alsace and Avignon madders to Mulhouse, and fertilizing it with chalk and horse dung. The results showed that control samples planted in the local soil without the additives produced weaker dyes. Madder grown in calciferous soil also dyed faster colors. It takes around three years for the roots to fully absorb calcium, which was generally considered the minimum period of growth needed.[77] A detailed study of madder roots tissue, structure, and contents was published in 1837 and became a formative text on the topic. Key observations were that the roots are initially yellow and darken with age, fresh plant material releases a yellow fluid that turns red in the atmosphere, and the fluid from older roots is darker yellow, becoming orange and red.[78]

As discussed earlier, plant species containing alizarin and varying mixtures of similar anthraquinones come from multiple genera, including *Rubia*, *Galium*, *Rebulnium*, *Morinda*, and *Oldenlandia*, the last of which does not contain purpurin, only alizarin.[79] The dyes in madder are

Figure 2.13 An 1846 cotton print made with madder on iron and aluminium mordants. The range of violet, red, and pink colors demonstrates the polygenetic property of madder. *Julie Wertz*

secondary metabolites, or compounds made by mature cells that have no clear role in cell growth.[80] Early research on madder root determined the yellow liquid secreted by fresh madder root, which was named rubian by English chemist Edward Schunck, does not contain free alizarin.[81] Rather, it is present in the form of compounds called glycosides, where the dye is bound to a sugar—in this case primeverose. In its bound form, alizarin primeveroside (the sugar-anthraquinone compound) is called ruberythric acid and the purpurin compound is called galiosin (which becomes pseudopurpurin first). After mature plants are harvested, enzymatic activity in the roots breaks, or hydrolyzes, the glycoside bond and frees the colorant from the sugar.[82] Research shows the alizarin content increases about 30 percent between two and three years of growth, a fact growers and dyers were clearly aware of before the chemistry of the roots was understood.[83]

The enzyme that hydrolyzes the glycoside bonds is called erythrozyme, and it is destroyed by boiling the roots in water. Soaking ground madder in warm water around 35–65 °C, however, was an efficient way to extract color by activating the enzyme. This likely explains why sun-dried Eastern Mediterranean madder gave better color than the kiln-dried Dutch roots, since higher temperatures deactivate more enzyme. Storing the madder in closed tubs to induce fermentation also increases erythrozyme activity, which explains why cask storage as practiced by the Dutch was an effective means to improve quality. Schunck found that the presence of calcium encouraged fermentation, which would also explain the superior coloring quality of madder grown in calcium-rich soil.[84]

Madder dyed beautiful and fast reds, but it contains many other substances in the roots that can dull or cloud the color, so obtaining a good result required a considerable amount of work.[85] The textile industry consumed enormous amounts of madder, and dyers and printers sought more efficient products both in terms of concentration and purity. In 1826, isolates of the main coloring components alizarin and purpurin were presented during a talk for the Académie des Sciences by French scientists Colin and Robiquet. They described experiments where madder was soaked in cold water for twenty-four hours, then boiled in new water, filtered, and treated with sulfuric acid, producing orange particles that were collected on a filter.[86] After Colin and Robiquet isolated alizarin and purpurin, experimentation began on modified madder products following their method of boiling and treatment with sulfuric acid.

The first commercial product, garancine, was introduced in 1830 by Lagier and Thomas of the firm Thomas frères in Avignon, and was also called "Lagier's bloom of madder." It had three times the dyeing power of madder, later improving to four or five times. About thirty kilograms of garancine could be made from one hundred kilograms of ground madder root. Garancine began with soaking ground madder in acidic water for six to twelve hours. For one hundred kilograms of madder, about one to two kilograms of acid and one thousand liters of water were used. The liquid was drained off and thirty kilograms of sulfuric acid or forty kilograms hydrochloric acid added to the wet madder paste. This was boiled for three to four hours, then filtered and washed with fresh water before pressing, drying, and milling. Garancine reds were not as fast as madder reds to washing or light, possibly related to residual sulfuric acid or side-reaction products. The treatment removed materials in the root that had a browning effect on the color, producing a clearer result. Another advantage was less bulk from a higher concentration of colorant, making it easier to store and transport. Dyers in Rouen and Alsace started to use garancine in the late 1830s. Archive material shows Scottish Turkey red manufacturers used garancine until the late 1870s and it is likely madder derivatives were employed generally in Turkey red production during the mid-nineteenth century. A garancine derivative, pincoffin, was made by heating garancine to 200 °C (390 °F) with steam, oil, or hot sand, and was particularly good for dyeing lilacs and violets. It was sold from 1854 by Simon Pincoffs, a Dutch chemist operating a garancine factory near Manchester.[87]

Another madder product called *fleur de garance*, or "flower of madder" was commercially available from Avignon manufacturers Julian and Roquer starting in 1851. Ground madder was soaked in cold water, which removed some water-soluble non-tinctorial components and induced fermentation of the glycosides. The mixture was treated with acid and placed in filters to ferment for five or six days, then the filter cakes pressed and dried in stoves before grinding back into a powder and packing in casks. Charles-Émile Kopp, who worked as chemist in an English Turkey red factory, researched madder glycosides and found that treatment with sulfuric acid followed by boiling in water freed a large quantity of alizarin by hydrolyzing the glycoside bond. He developed a preparation called "Green Alizarin," which was made by taking the acidic wash water from a garancine batch and adding more acid, then heating it to 60 °C (140 °F). This formed an orange precipitate that was collected on a filter and washed. It was composed of nearly pure purpurin, which could be further purified with other treatments.[88] Green Alizarin was also known as "Kopp's purpurin" and was used in France to make pigments for paint until the early twentieth century.[89] Anthraquinone glycosides are more water-soluble than the free dye, so Kopp's preparation was recovering still-dissolved colorant by completing the hydrolysis and precipitating the dye. Since the final product is purpurin, the compound in the liquor was probably galiosin.[90]

Synthetic Alizarin

The 1856 synthesis of mauveine by William Henry Perkin was significant because of its potential to free the textile industry from dependence on the vagaries of a natural products, whose quality and yield varied. The ability to manufacture dyes from coal tar, the material he started with, has been called "one of the best and most unexpected discoveries of contemporary chemistry." The supply of coal tar was a by-product of the gas lighting infrastructure that first appeared in Britain in the early nineteenth century. Experiments starting in the mid-seventeenth century had shown that distilled (rather than combusted) coal released gas, a mixture of volatile carbon-based fuels that could be burned for illumination. In 1802, Scottish engineer William Murdoch used this "inflammable air" to illuminate parts of his employers' factory, the Birmingham firm Boulton and Watt. They took the system to market, and the cotton mill of Philips and Lee in Salford, near Manchester, was the site of the first industrial application of gas lighting. Street lighting was installed in parts of London in 1807 and expanded rapidly from there. Since the coal did not undergo combustion, the heavier compounds remained, forming the coal tar residue from which synthetic dyes were made.[91] Coal tar contains a complex mixture of hydrocarbons, many of which were involved in dye synthesis, and are often collectively referred to as "aniline" dyes after one particularly significant component ($C_6H_5NH_2$).

Understanding Alizarin

Following mauveine, there was immediate interest in synthesizing alizarin to reduce dependence on madder. Although it was the first dye for which there was a significant understanding of its structure, which facilitated its eventual synthesis in 1868, it took some time to achieve this knowledge.[92] Colin and Robiquet had isolated alizarin in 1826, but they were not able to fully describe the molecule. At the time, experiments to determine molecular structure involved breaking the compound down to see what product(s) remained using solvents and heat. Schunck found that alizarin extracted from madder, then treated with nitric acid yielded phthalic acid, which was a decomposition product of naphthalene($C_{10}H_8$, see Figure 2.14). This caused a misconception that alizarin had a two-ring structure, rather than the three-ring anthraquinone, and some supposed that converting naphthalene could be a synthetic route for alizarin.[93] Around 1850, German chemists Strecker and Wolff gave the chemical formula of alizarin as $C_{10}H_6O_3$, which is based on a naphthalene structure. Schunck disagreed with this conclusion and gave the formula $C_{14}H_{10}O_4$, which is much closer to the actual $C_{14}H_8O_4$, but this does not appear to have had much influence on the consensus at the time. The confusion persisted for eighteen years, hampering synthesis.[94]

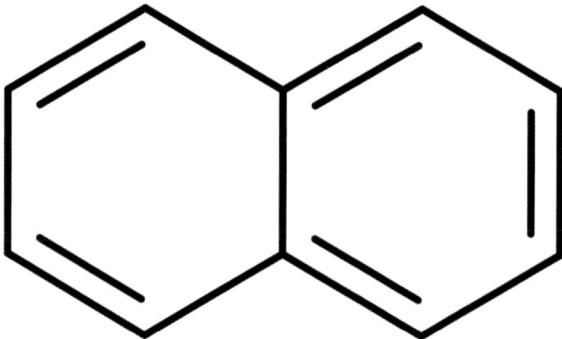

Figure 2.14 Structure of naphthalene.

Eventually, further research clarified the structure of alizarin and it became the first natural dye replicated in a laboratory. The breakthrough was made by another pair of German chemists, Charles Graebe and Charles Liebermann. Graebe was an assistant of Adolf Baeyer, an accomplished German chemist who had studied under Robert Bunsen, inventor of the Bunsen burner. In 1865, Baeyer was a professor at the Gewerbe Institut in Berlin where he hired Graebe, also a former Bunsen student, who had studied anthraquinones. Liebermann, a former student of Baeyer, was hired in late 1867 after working for two years in dyeing, first at the Turkey red works of Koechlin, Baumgartner & Co. in Loerrach, Germany (near Mulhouse) and then his father's calico printing factory in Berlin. The combination of Graebe's knowledge of anthraquinones and Liebermann's practical skill with madder products was well-suited for their work on synthetic alizarin, which began in early 1868.

Through experiments to break down the molecule, they determined its base structure was anthracene with fourteen carbons, rather than the ten carbons of naphthalene (see Figure 2.15). This was achieved by distilling extracted alizarin with zinc dust. Both anthracene and naphthalene break down into phthalic acid, however, which explains the original discrepancy between the theory of Schunck and that of Strecker and Wolff. Graebe and Liebermann determined that alizarin is an anthraquinone with two hydroxyl groups, and gave its correct formula, though it took a little longer to determine the rings in a linear arrangement.[95]

Alizarin Synthesis and Patent Disputes

In the summer of 1868, Graebe and Liebermann began attempts to synthesize alizarin from anthracene, which meant finding a means to oxidize it to anthraquinone (Figure 2.11) and then hydroxylate it to alizarin (Figure 2.12). These reactions could also be undertaken in the opposite order (hydroxylation and then oxidation), and under different conditions, with varying degrees of difficulty. Anthracene was not commonly available at the time since it had no use, but Graebe and Liebermann obtained enough to experiment by oxidizing it with nitric acid. Hydroxylation was a more difficult reaction, so they pursued an indirect route via anthraquinone sulfonation, a recently discovered reaction where sulfonic acid groups were added and then converted to hydroxyls. They

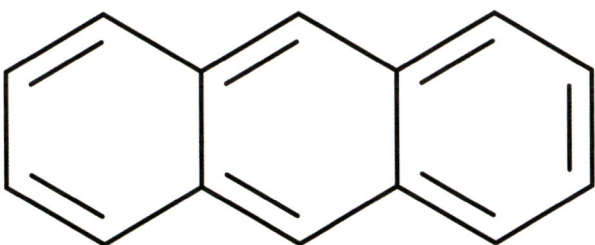

Figure 2.15 Structure of anthracene.

Figure 2.16 A small amount of orange-red synthetic alizarin powder can replace a much larger quantity of ground madder in dyeing. It contains no woody impurities, but will produce a slightly different shade of red. *Julie Wertz*

were unsuccessful, and found another alternate route by adding bromine to anthraquinone, then replacing the bromines with hydroxyls. Figure 2.16 shows a few grams of modern, nearly pure synthetic alizarin, a fine reddish-orange powder with high tinctorial power.

Graebe and Liebermann reported their synthesis at a meeting of the Chemische Gesellschaft, the precursor to the German Chemical Society, on January 11, 1869. They brought samples of their synthetic alizarin and textiles swatches dyed with it. The chemists secured a patent for their method and signed a contract with Badische Anilin und Soda Fabrik (BASF), which had been founded only a few years earlier by Friedrich Engelhorn and manufactured synthetic dyes and chemicals for the dye industry. Engelhorn had built a gas works in Mannheim in 1861, which conveniently provided a steady supply of coal tar, and the firm had accumulated a large supply of unused anthraquinone that would become valuable if converted into alizarin.

One of the persistent challenges in industrial chemistry is translating laboratory-scale processes to production scale. Sufficient mixing, even heating, and whether heat or gases are produced are among the various factors that must be accommodated and adapted for. Heinrich Caro, a BASF chemist, was tasked with helping Graebe and Liebermann adapt their patented bromination method for production, which proved difficult. Although successful in the laboratory, the route was not commercially viable due to its low yield and the high cost of bromine. Caro had been hired at BASF in 1868, shortly before they acquired the alizarin patent. He had studied textile dyeing at the Gewerbe Institut, then went to England to learn about the industry in Manchester. During his time there, he also visited Schunck, the madder expert, and worked in aniline dye manufacturing. The trio returned to pursing anthraquinone sulfonation, which Graebe and Liebermann had abandoned. Caro accomplished it by accident when he left a reaction mixture of anthracene and sulfuric acid heating while away from the bench. Upon his return, the laboratory was filled with dense fumes and the mixture was nearly dry, but a pink crust had formed around the edge of the dish. Caro realized it was alizarin, and that sulfonating anthraquinone required higher temperatures and stronger acid than Graebe and Liebermann had previously used. Experiments showed that heating anthraquinone with concentrated sulfuric acid to at least 200 °C (390 °F) made anthraquinone sulfonic acid. This was heated to 180 °C (350 °F) with sodium hydroxide to complete the synthesis. In June 1869, Caro, Graebe and Liebermann filed new patents in Germany and Britain.

Graebe and Liebermann were not the only chemists looking for a means to sulfonate anthraquinone since their announcement in January of 1869. As a student at the Royal College of Chemistry, Perkin had worked on anthracene compounds for his supervisor August Hofmann. He retrieved his samples and began investigating the sulfonation route, producing alizarin in May of 1869. He dyed some samples and sent them to a friend, Robert Hogg, who was a merchant to the Glasgow dyeing industry and dealt in Turkey red. Hogg was impressed with the result, so Perkin filed a British patent on June 26, 1869.

Meanwhile, the German chemists were disappointed twice over. Although the sulfonation method was suitable for industrial production, domestic patent law did not grant it separate protection on the basis that it was too similar to the bromine method. This, however, did not mean they also acquired protection under the bromine patent, and BASF was left with exclusive rights for an unworkable process while competitors were free to make alizarin via the unprotected sulfonation route.[96] Their British patent application was filed on June 25, 1869, coincidentally the day before Perkin filed his. In a potential show of national favoritism, Perkin was granted a patent on August 24 of that year (GB1869 1948).[97] The Germans' application was delayed for some weeks due to a request for rewording, and not granted until January 11, 1870 (GB1869 1936).[98]

This left Perkin with the unquestionable advantage of patent protection for a viable process, coupled with access to a large domestic textile industry that consumed vast quantities of imported madder. He was limited, however, by the need to import oleum, a highly reactive mixture of sulfuric acid and sulfur trioxide that was more effective than sulfuric acid alone for alizarin synthesis. Oleum was imported in ceramic bottles from Germany, and was expensive due to the distance and hazards of transport. Rather than accept the situation, Perkin continued to research alternate synthetic routes to avoid importing it.[99] He developed a route via chlorination, where purified anthracene was spread on trays placed in lead ovens, which were then filled with toxic and highly reactive chlorine gas. This was made by mixing hydrochloric acid with calcium hypochlorite. The anthracene became chlorinated over five to six hours in the oven, yielding hydrochloric acid gas as a by-product that was drafted out chimneys along with the excess chlorine gas. Most of the anthracene became dichloroanthracene, which could be converted to anthraquinone sulfonic acid with sulfuric acid, thereby avoiding the need for oleum. The final hydroxylation step was the same as the sulfonation process.[100] Perkin filed a British patent application for this method on November 17, 1869, and it was granted on January 7, 1870 (GB1869 3318), so his second patent was sealed four days before the Germans' first one.[101]

There are four synthetic pathways to make alizarin from anthracene in the early patent literature—one via bromination and three via anthraquinone sulfonic acids. Only the two routes for which Perkin held British patents were commercially relevant (see Figure 2.17). The most common industrial route, which is still in use today, was to oxidize anthracene to anthraquinone and sulfonate it with oleum.[102] Although there are few documented descriptions of the synthesis, there is enough information to give a sense of what it was like. Pure anthracene is a pale greenish white, granular solid. The initial mixture with sulfuric acid turns bright yellow, then quickly becomes a thick, greenish-brown sludge as "the consistency of treacle."[103] After washing and neutralization, the anthraquinone sulfonic acids were collected as pale gray flakes sometimes called "silver salts."[104] During hydroxylation, the silver salts become a thick, blackish mass that turns a brilliant amethyst purple when diluted in water. Dropping acid into this mixture precipitates the dye, an orange-red solid, which is collected on a filter.

Despite their significant contributions to the synthesis of alizarin, the German chemists were left without patent protection in any market. Rather than pursue legal action, Caro, Graebe, and Engelhorn met with Perkin in England to seek an agreement to divide the market.[105] Perkin helped the Germans revise their British patent to mutual satisfaction, and in March 1870 the two parties ratified an agreement where Perkin held monopoly over the British market and the Germans took the European mainland. Unfortunately again for the Germans, the outbreak of the Franco-Prussian War that year limited production, and without patent protection BASF still had domestic competition.[106] Perkin enjoyed a short monopoly before the war ended and the British market was flooded with imported alizarin products made by these foreign competitors. They also courted representatives from Scottish Turkey red firms, hosting visits and lavish parties to gain their business. The arrangement suited both parties, since neither had agreed to the cartel formed

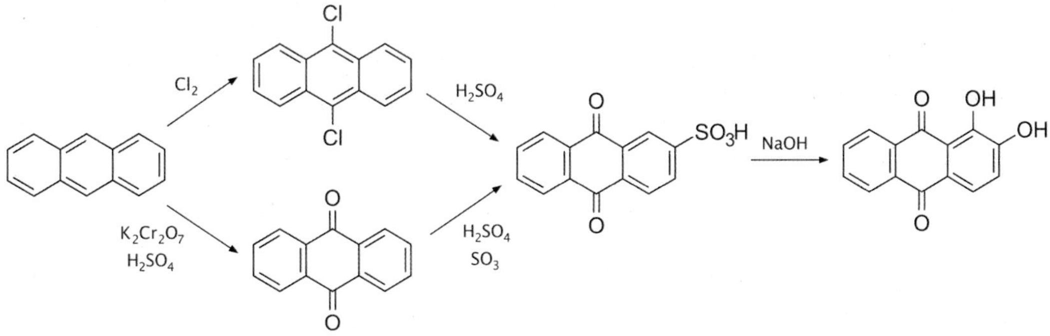

Figure 2.17 Synthetic routes for commercial alizarin production in the late nineteenth century. The raw material anthracene (left) could be chlorinated (top) or oxidized (bottom), then sulfonated to anthraquinone sulfonic acids before hydroxylation.

between Perkin and BASF. To circumvent any potential legal issues, imports were intentionally mislabeled as madder or garancine.[107]

Synthesizing alizarin on the desired scale required a supply of sufficiently pure anthracene. Since it was not yet commercially available, Perkin implemented in-house coal tar distillation at his factory, Greenford Green, a process he had learned from his experiments as a student. Eventually, his partner and brother, Thomas Dix Perkin, showed local tar distillers how to do some of the work, but their anthracene still required further purification. Impurities could waste reagents, form undesirable products, and hinder filtration during manufacture. The yield of purified anthracene from coal tar was about 1 percent of the original mass of coal tar.[108]

Synthetic alizarin was adopted gradually by dyers since the early product was not immediately of a sufficient quality to totally replace madder. Throughout the 1870s, manufacturers sought to optimize their processes and eliminate technical problems like extraneous coloring material and variable quality. Synthetic alizarin was first sold at 10 to 15 percent concentration, later 20 percent. It came in a paste form that was easier to dissolve in water than dry powder, which tended to leave spots on the final product due to the difficulty of dispersing it evenly. Later, some pastes were sold at concentrations up to 40 percent. The cost of dyeing with synthetic alizarin became a fraction of what it cost to use madder, causing an irreversible shift away from using it. In France, madder prices dropped about 60 percent in an effort to preserve some market share. The Société d'agriculture de Vaucluse, concerned about the future of madder, made formal enquiries to the Mulhouse textile industry about using artificial alizarin, whether it dyed and printed as well as madder, what the price was, and, ominously, what madder growers should fear for the present and future.[109] In Smyrna, the lack of demand caused madder crops to be left in the ground and farmers offered payment to anyone willing to dig it up to free the land. Despite the economic devastation for madder growers, the adoption of synthetic alizarin and subsequent disuse of madder freed significant amounts of arable land to grow food crops instead.[110] Within five years, the Turkey red industry had largely abandoned madder, an industry once worth millions.[111]

Increasingly efficient processes lowered the price of synthetic alizarin to the point where manufacturers sought a cooperative system of price and production control—another cartel. In 1870, a pound of 10 percent paste cost six to seven shillings (about £20 today), and by 1900 a pound of 20 percent paste was ten pence (just over £3 today). On the US market, it was $3.37 per pound in 1872 and $0.16 in 1890.[112] In 1881, all major European producers agreed to a cartel that parceled out trade throughout Europe and gave the United States to BASF. Called the Alizarine Convention, it included agreements on minimum pricing, production quotas, and recognition of a control board established to monitor compliance and handle disputes.[113]

Synthetic Alizarin Products

Experiments on the new synthetic alizarin revealed it was not pure, much like how madder also contains mixed anthraquinone colorants. Experienced dyers could see the difference in test batches of Turkey red dyed with synthetic alizarin and with madder, which were slightly different shades of red. Although chemists knew the structure of alizarin and multiple ways to manufacture it, they did not have a thorough understanding of how the reaction proceeded and what intermediate and side products formed. In their original bromination method, Graebe and Liebermann incorrectly determined where the bromine attached to the anthracene carbons. The same uncertainty existed regarding the sulfonation process. Initially, they thought anthraquinone acquired two sulfonic acid groups, but further research revealed that alizarin came from anthraquinone with a single sulfonic acid as shown in Figure 2.17. When the reaction proceeded further to yield disulfonic acids, they were in fact making novel dyes.[114]

Gustav Auerbach, a German chemist, was the first to isolate one of these compounds. He determined it was an isomer (same formula, different structure) of purpurin, and called it isopurpurin. Perkin also isolated the dye but called it anthrapurpurin, which is the name used today. He found it produced brighter shades of red than alizarin, with slightly lower wash and light fastness. The other novel colorant, another trihydroxyanthraquinone isomer called flavopurpurin, was identified shortly afterward (see Figure 2.18). Neither anthrapurpurin nor flavopurpurin occur in madder. One recent study cites flavopurpurin in extracts from *Rubia tinctorum*, but it is a review that provides no analytical data. No other analyses of madder have found either dye present. Synthetic alizarin was expected to also contain purpurin, but surprisingly it was not found. Further research determined purpurin is synthesized through a separate reaction between alizarin and sulfuric acid with manganese peroxide or arsenic acid, a method developed some years later. This was more expensive than alizarin synthesis, and probably not carried out on a large scale since it is rarely mentioned in the literature.[115]

By 1878, most issues were sufficiently resolved, and production was efficient and profitable, yielding a satisfactory amount of useable dye. Chemists determined that lower temperatures and shorter reaction times during anthraquinone sulfonation yielded the monosulfonic acid that became alizarin, while higher temperatures for a longer time made disulfonic acids that yielded anthrapurpurin and flavopurpurin. This knowledge made it possible to better control the product by manipulating the reaction conditions. The variance in red hues that different relative amounts of these dyes produced led to the names "red" or "scarlet sulfonation" (more anthrapurpurin and flavopurpurin) and "blue sulfonation" (more alizarin) for the tint they dyed. Perkin's chlorination route produced dye with a higher concentration of anthrapurpurin than the alizarin, more like the sulfonation "red shade." It became common to sell synthetic alizarin under various color codes that

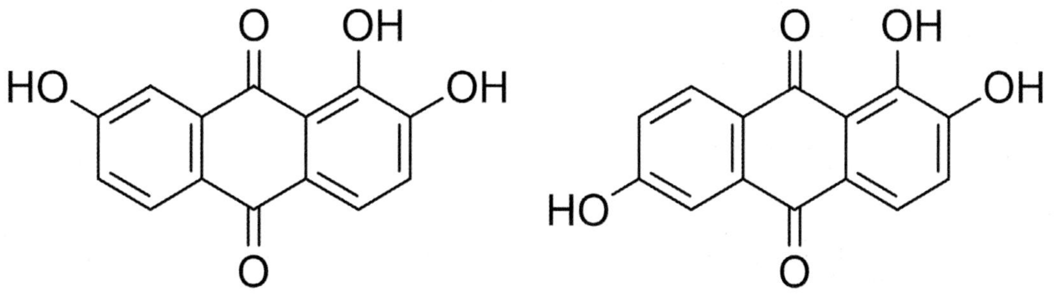

Figure 2.18 Structure of anthrapurpurin (left) and flavopurpurin (right).

would indicate the resulting shade, which depended on the ratios of alizarin, anthrapurpurin, and flavopurpurin. The number of codes was extensive (e.g., R, RA, GA, V, VX, SX, SGX), and effectively worthless.[116] While it was useful for consistency to purchase the same blend from a manufacturer, there is no indication coding was standardized across manufacturers. Furthermore, the limited control over synthesis and purification, and there only being three major colorants present, means in practice the number of combinations implied by the lengthy code list would have functioned more as a marketing gimmick. Taking advantage of available resources, some dyers mixed the anthrapurpurin-heavy "red shade" of alizarin with garancine to balance the color. This fell out of practice as madder and garancine disappeared from the market and better control over alizarin synthesis yielded more consistent product. Not all firms were enthusiastic about the switch, however. Prévinaire of Haarlem continued to use garancine until it was no longer manufactured in sufficient quantities and their backstock had been exhausted.[117]

The presence of anthrapurpurin and flavopurpurin on a piece of Turkey red indicates it was made no earlier than 1870, based on the history of synthetic alizarin and their function as chemical markers for its use. Purpurin indicates a madder or madder derivative was used, since synthetic purpurin was never a commercial product, and may be accompanied by xanthopurpurin, pseudopurpurin, and other minor components. This profile does not guarantee a textile was made before 1870, since madder products continued to be used for a time, but it is unlikely Turkey red dyed with madder was made after the 1880s. The transition from madder to synthetic alizarin was a significant one for the textile industry and industrial chemistry, and was fueled by the production of Turkey red. The use of these dyes is discussed further in Chapter 3.

Conclusion

The history and origins of Turkey red are explored here using its reputation as a fast and distinctive color on cotton, following the material- and process-based definition given in Chapter 1. This enables comparison across time and cultures, where different materials are used to fulfill the same chemical function in the dyeing. While somewhat speculative, the lack of detailed records and surviving objects pre-dating the eighteenth century requires some indirect study. This would not be possible in most cases, but Turkey red appears to be exceptional in its unique status as the most brilliant red dyed on cotton prior to the early twentieth century, giving more confidence to the assumption that mentions of red cotton refer to Turkey red.

The exchange of knowledge and application of emerging scientific concepts are two significant aspects of Turkey red history, which likely originated in India. There is evidence of dyeing practice in multiple regions, particularly Gujarat and the southeast coast, that is consistent with the definition used here, and centuries of documentary evidence for red cotton being traded from India. Similar practice is also found in Indonesia, including using aluminium accumulating plants, and of course in the Levant, where its European name comes from, indicating knowledge spread east and west from India. Ottoman dyers and traders facilitated Turkey red trade further westward into Europe, and in the eighteenth century aspiring French and English dyers sent agents to the

Eastern Mediterranean to gain practical knowledge of the process. The Industrial Revolution facilitated Turkey red dyeing in Western Europe, which was finally established there through a combination of immigrant dyers and industrial espionage, quickly outproducing the industry in India and the Levant. This eroded local markets and reversed the flow of trade as colonial powers were established.

Madder, the plant source of the red dye in Turkey red that was used for most of its history, was itself a significant part of the textile industry and a major trade commodity. It was also a subject of interest for pioneering chemists studying natural materials. By the mid-nineteenth century when the potential for synthetic dyes was discovered, the Turkey red industry was consuming enormous quantities of madder. Chemists sought to understand the structure of its main colorant alizarin, and a means to replicate it synthetically. Commercial alizarin became the first natural dye replicated synthetically, and replaced madder in Turkey red dyeing by the 1880s.

The Dyeing, Chemistry, and Technological Advances of Turkey Red

Chapter Outline

There are two eras in the history of Turkey red dyeing, often called the "old" process and the "new" process. The distinction emerged in the early 1870s with the new process, which used a recently developed product called Turkey red oil instead of rancid plant oils. Significantly, this reduced the overall duration of Turkey red dyeing from several weeks to about three days.[1] Both processes yielded cotton imbued with fatty acids that was ready for the subsequent treatments, but the oil application was markedly different. Around the same time, synthetic alizarin began to replace madder in the dye bath. This transition is not included in the terms old and new process, but certainly would have contributed to the increased efficiency since madder dyeing required more washing than synthetic alizarin.

Chemistry emerged as a distinct field separate from alchemical theory in the mid-seventeenth century, so for as long as Turkey red dyeing was practiced in Western Europe, curious minds sought to understand the process. Although they did not have the level of scientific capability available today, early chemists were skilled at identifying materials. Dyers recognized that they were following a chemical process which benefited from research, and understanding their work was the best way to improve it.[2] The first Turkey red methods were published in the mid-eighteenth century, followed closely by investigations and theories on the chemistry of the process.

This chapter discusses the old and new processes of Turkey red dyeing stepwise. It also explores early research on the chemistry of Turkey red and what is now known since improvements in chemical knowledge and analytical techniques finally helped explain how the color forms. Written records of dyeing Turkey red in English and French from the mid-eighteenth to early twentieth centuries were surveyed. Some include directions for workshop setup, vessel composition, commentary on the quality of water, and advice for handling techniques.[3] Others are less descriptive and may leave the modern dyer with questions about what exactly to do. It was known that "almost all dyers of Turkey-red follow processes of their own invention, and keep these modes of procedure strictly secret."[4] Studying a range of methods enables broader observations about what is representative and what is atypical, making it possible to infer more about the chemistry of Turkey red in addition to helping define the process. This level of access to knowledge was impossible prior to the digital era since many of the references to Turkey red are dispersed throughout literature on other topics. Historical texts state that a certain amount of practical knowledge was needed to successfully dye Turkey red, and that even successful results had a variable quality, especially regarding fastness. This is explained mostly by a dyer's practical knowledge and available ingredient quality, which also required knowing how to recognize good materials. Although using an excess of ingredients increased chances of success, for commercial viability a method had to be optimized.[5] The knowledge and skill developed from this work facilitated the innovations that lead to the new process.

Oiling

Cotton fibers must be scoured before dyeing Turkey red, as they do for all dyeing and printing, but do not need to be bleached. Dyeing started with the oil treatment, which was called "the chief factor in the whole process," and had to be done with care to ensure a good product.[6] Oiling was the foundation of Turkey red, and the final outcome of the color and quality depends heavily on its success, which is determined by ingredient quality and the experience of the dyer. Otherwise, the color was poor and uneven.[7] Studies have shown that more oil treatments yield a deeper shade of red, a direct correlation emphasizing its importance to Turkey red.[8] In texts, the oil is often called a mordant or an assistant. Since an assistant facilitates the process but is not incorporated into the final product, while the oil was, mordant is the more accurate designation for it. The application was lengthy, laborious, and likely the greatest cause of error in failed batches. Furthermore, since it is particular to Turkey red (oils are used in other textile processes, but not in this manner), those learning the process did not have experience to rely on.

Although its importance was known, the exact purpose of oiling was not fully understood during the era of Turkey red manufacturing. Experimental work on its role included the observation that oil dissolved the color from finished Turkey red, indicating it likely contained oil itself.[9] While oils were reasonably well characterized in the nineteenth century, cellulose was not until the early twentieth century, limiting chemists' ability to explain interactions between the two. Since cotton fibers are integral to dyeing Turkey red, an incomplete understanding of its chemistry prior to the

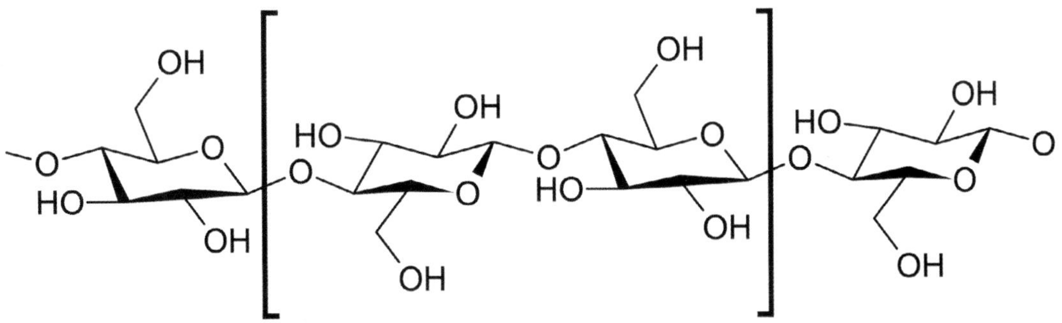

Figure 3.1 Cellulose structure with brackets around the segment that repeats to form the polymer.

1930s limited early chemists' efforts to understand the color, though attempts were made to do so.[10] Materially, cotton is the purest form of cellulose found in nature. Cellulose is a linear polymer, or polysaccharide, of glucose rings linked by covalent bonds (see Figure 3.1).[11] The hydroxyl groups on the rings are where the color complex is bonded, discussed further in the following sections. Linen is also chemically comprised of cellulose, but with a more rigid structure that makes it more difficult to dye with natural dyes.[12]

Oiling in the Old Process

The key property of oil used in the old process is a high free fatty acid content. Chemically, oils are triglycerides, or three fatty acids joined to a glycerin molecule (Figure 3.2). When the bonds with glycerol are broken, the oil undergoes a process called hydrolytic rancidity and releases fatty acids (Figure 3.3). As such, the oil treatment is not materially an oil, but rather a mixture of fatty compounds including partial glycerides and free fatty acids. Chemists knew that a good dyeing oil had a high fatty acid content, and researched the processes of triglyceride breakdown by

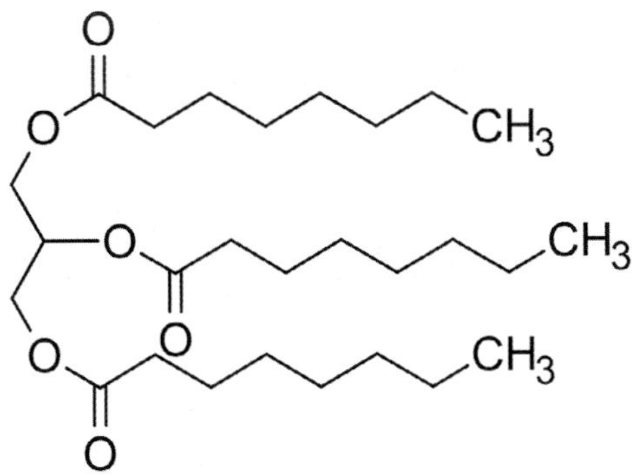

Figure 3.2 Triglyceride molecule showing three fatty acids bound to glycerol.

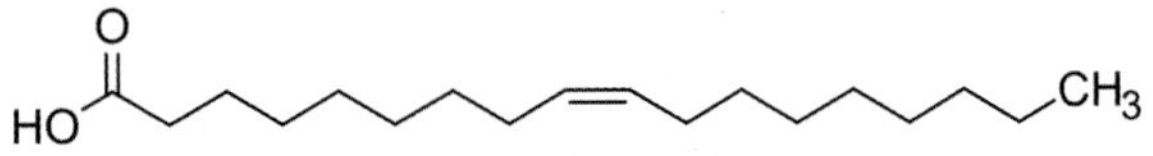

Figure 3.3 Structure of oleic acid, a fatty acid.

fermentation, oxidation, heating, boiling with acid, and other manipulations.[13] Grease stains in an "unmodified state," or intact triglycerides, formed a resist on the fiber and prevented metal mordants from attaching.[14]

The dyeing practices discussed in Chapter 2, which begin with a rancid oil treatment, are characteristic of old process methods. The source varied with geography and local availability, but the common feature was an elevated fatty acid content. In India, dyers used buffalo or ewe's milk, or Illipé butter (*Shorea stenoptera*) as the fat source. Sesame oil was also an option in India and Türkiye. Lard is mentioned, but solid fats like this and Illipé butter were probably more difficult to evenly distribute on the fibers. On the Caspian Sea coast, Armenian dyers used fish oil from nearby fisheries, primarily sourced from sturgeon and shad. Traditional dyeing still practiced in Indonesia include oils from the plants *Pangium edule*, *Sterculia foetida*, *Schleichera oleosa*, and *Aleurites moluccana*. In Europe, the best oil for dyeing was the rancid olive oil often called Gallipoli oil (for the Italian seaport) or *huile tournante* ("oil on the turn"). The first cold pressing of olives yields extra virgin oil, while second and third pressings with boiling water and fermentation extract an inedible oil used for soap making and textile dyeing. Gallipoli oil contained a lot of mucilage, the particulate matter from the crushed olive fruit, which was considered useful. Mucilage facilitated rancidification with its content of sugars and sugar acids, which are monosaccharides with at least one terminal carboxylic acid group. Acids in general, as discussed later, improved the quality of oil for dyeing to a point. Sometimes fatty acids, like oleic or margaric acid by-products from making tallow candles, were also added to improve the quality. Colza (rapeseed, *Brassica napus* subsp. *napus*) oil was used in Germany for a period in the early nineteenth century, but did not produce good results and was not much used in Turkey red. Although cheaper, oils like colza, poppy, or palm were more stable, and failed to provide the necessary fatty acid content. Considering the importance of the oiling treatment to a successful Turkey red batch, and its unique status as a pre-treatment in the process, it probably took new dyers some time to develop knowledge of what constituted a good oil. Quality was not always explicit in publications. For example, an 1871 method in *Scientific American* says to use olive oil, without mentioning rancidity.[15] Since this was not a specialist publication, it is unlikely a reader would already know that important detail.

To apply the treatment, the oil was mixed with an aqueous alkaline solution often called a "ley" or sometimes "lixivium." Leys were made by dissolving soda (sodium carbonate) in water. From the nineteenth century, soda was manufactured through a chemical process. Prior to this, soda was made by burning plants and wood and collecting the ashes. Eighteenth-century French and English dyeing texts say that best quality soda came from Alicante, on the Mediterranean coast of Spain, where it was made by burning *barilla*, or saltwort, a halophyte plant that grows well in saline environments. The soda was also called Alicante soda, *kalakar*, or *barilla salsola*. A good test of oil quality was to mix a portion with some ley, which should form an emulsion that looked like "a dense milky fluid" and did not separate or become curdy after twenty-four hours. Because of its appearance, this oil treatment was often called the "white bath." An older term for it is *sikiou*, which is first documented in French around the same time the first Turkey red works were established there, and may be a transliteration from a Greek term brought by migrant dyers. One method suggests adding a fresh egg yolk to the oil bath if the emulsion was not persistent. This was not widely practiced, and probably only functioned as an emulsifier in the bath, effectively masking the quality if the fatty acid content was too low. A French dyer, Jean-Michel Haussmann, promoted linseed oil for Turkey red, claiming it made a better product. He used strong alkalis, boiling water, alum, and linseed oil in his bath, which formed the required "milky liquid," albeit one that separated and had to be shaken before use, indicating a lack of stability.[16] Furthermore, the dyer should not add aluminium to the oil treatment since it will interfere with the chemistry of the process by preferentially binding with the oil over the fibers. Other dyers, skeptical of his method, attempted to re-create it with little success, and Haussmann never produced a significant amount because it scaled up poorly. In fact, linseed and other types of drying oils, which react with the atmosphere to cross-link and form polymer structures, were generally discouraged in Turkey red for the blackening effect they had on the final color.[17]

The cotton fibers (either yarn or cloth) were saturated with this rancid oil and alkaline ley emulsion before drying, typically three times, then rinsed to remove excess oil. This was repeated to obtain a higher-quality textile with more saturated color. As a result, an old process method could have around sixteen steps, many of them repeated oil applications and washings. One method from the early nineteenth century was said to take about sixteen days to complete; later publications claim a month was needed. A single treatment involved steeping (often referred to as "tramping") the cotton in the oil bath to saturate it, possibly leaving it to soak for twelve to twenty-four hours. Some establishments warmed the oil baths to about 28 °C (85 °F), but most methods do not specify a temperature. Some dyers worked with vessels of hollowed-out wooden blocks, steeping the yarn by pouring over a shallow portion of warm solution rather than maintaining a warm immersion bath. An early dyeing text recommended using vessels made of white softwood with a low tannin content, like pine or fir. Another method recommends copper vessels or wooden tubs, but only if hooped with wood or copper and without any iron nails or banding. Iron was cautioned against, even for steam pipes, since it could drastically dull the final color. Nevertheless, in the mid-nineteenth century some dye works used cast-iron vessels, so it must not have been problematic in all circumstances. After application or soaking, the excess liquid was removed by hydraulic press since wringing weaken the wet fibers and tangle skeins. A considerable amount of physical work was required to work wet cotton in solution, then lift, drain, and press it. Eventually, the application was mechanized to facilitate continuous application and wringing (see Figure 3.4). Some methods instructed the dyer to save unused oil bath and the wrung excess for use in the final clearing step, which is discussed later.[18]

Thorough drying improved the quality and color of Turkey red. It was essential to dry the fibers between every treatment, and required some care to execute properly.[19] Yarn hung outdoors to dry was brought indoors overnight if there was risk of dewfall wetting the oiled fibers, and after steaming was introduced to the process it was necessary to prevent condensation dripping on the cotton.[20] In some places in continental Europe, open-air wooden towers were built to facilitate drying textiles (see Figure 3.5). Some methods direct the dyer to leave the cotton in wet heaps for around half a day to induce fermentation before drying, which, under the right conditions, could harbor maggots.[21] This heaped, oily cotton also generated heat that presented a combustion risk, so it had to be monitored to avoid loss of product or the entire establishment.[22]

Stove drying was recommended when air drying was not possible due to climate conditions.[23] The stove was a heated, closed room sometimes also called a stoving room (see Figure 3.6).[24] Turkey

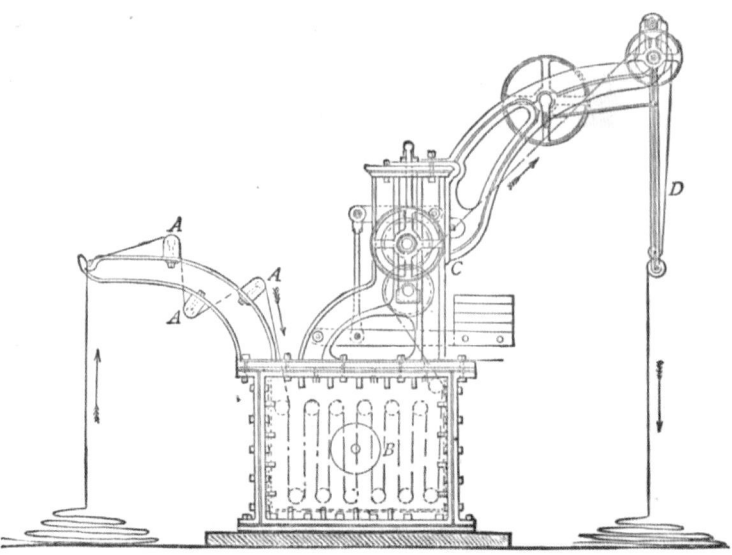

Fig. 91.—Oil-padding Machine.

Figure 3.4 Machine for application of oil treatment, late nineteenth century. *The Dyeing of Textile Fabrics. Hummel, John James. Cassell & Co., London. 1885*

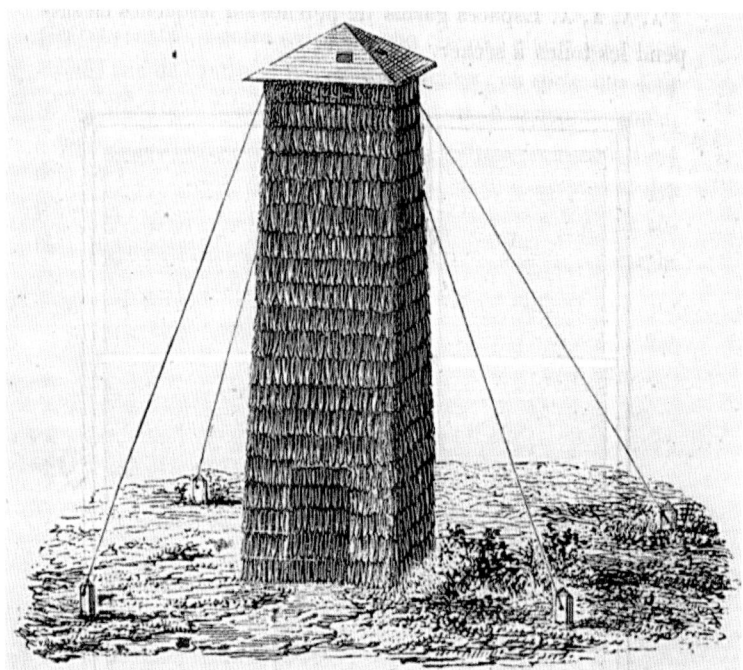

Figure 3.5 Wooden tower for drying textiles during dyeing, mid-nineteenth century. *Traite theorique et pratique de l'impression des tissus, volume 2. Persoz, Jean-Francois. Victor Masson, Paris. 1846*

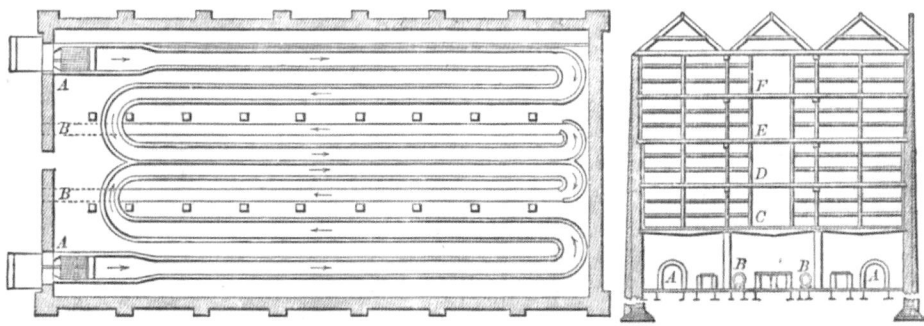

Fig. 93.—Ground Plan and Sectional Elevation of Turkey-red Stove,

Figure 3.6 Drawing of stove apparatus for Turkey red dyeing by Hummel. *The Dyeing of Textile Fabrics. Hummel, John James. Cassell & Co., London. 1885*

red dyers in Rouen were early adopters of stove drying since the climate was often cold and damp, whereas dyers in warmer Provence could rely on air drying. In Scotland, Turkey red was seasonal work and stopped during the worst winter weather until the early nineteenth century when stove drying was adopted.[25] The best temperature was around 65 °C (150 °F).[26] Sometimes the oiled cotton was air dried for a period, then finished in the stove.[27]

It was also important to remove excess oil from the cotton since not all of the treatment bonds to the fiber and loose, excess oil interferes with subsequent steps.[28] The dry, oil-imbued cotton was washed in a ley (sometimes warmed), and perhaps left to soak for a few hours. It was important to not wash so thoroughly as to remove the bound oil, weakening the color. This cleansing step was sometimes called degreasing (*dégraissage*) or salts (*sels*), and, though less often, "white" bath since the ley acquired a milky appearance as it dissolved excess oil.[29] Multiple washes were needed to ensure the excess oil had been successfully removed. Afterward, the cotton was washed in clear water before drying again. Not all methods clearly communicated the significance of this step, and it is unknown whether the omission was an intentional effort to restrict imitation, assumed to be tacitly understood, or poor communication.[30] Pinks, effectively light reds, resulted from insufficient oil, improper application, or excessive cleaning.[31] This was eventually exploited in a controlled manner to intentionally make shades of pink (*roses*) with less oil and less of the following aluminium treatment.[32]

The Chemistry of Oiled Cotton

In the early nineteenth century, a theory emerged that the oil had both a physical and chemical function.[33] It was observed that shredded Turkey red fibers had a red exterior and white interior, leading to the conclusion that the color was superficial on the fiber, possibly due to the oil, which both attached the color and protected the fiber like a varnish.[34] A minority opinion was that the oil had the opposite effect and "opened" the cotton fiber, causing it to take more aluminium, but did not remain on the cotton itself.[35] Systematic testing showed cotton increased in weight following oil treatments, and extractions of oiled cotton with organic solvents removed a thick, fatty material from the fibers, evidence that something was adhering to the cotton.[36] Although well-informed, chemists of the nineteenth century did not have the knowledge or resources to prove that the oil had a dual physical-chemical function in Turkey red, or to explain exactly how it worked. Analysing the matter extracted from oiled cotton showed what was present, but did not reveal how it bonded, and they could not analyse it in place on the fiber. The proposed chemistry was that the emulsion of alkaline ley and rancid oil formed a soap solution, which saturated the fibers.[37] Soaps are the metal salts of a higher molecular weight fatty acid like palmitate, stearate, and oleate, and a good oil bath had a higher concentration of free fatty acids. Experienced dyers said that gradual drying produced the best results by allowing more of the treatment to adhere to the cotton, indicating the uptake was slow and benefited from the presence of some water.[38]

Chemists interpreted the heat generated by heaped, oiled cotton as evidence of oxidation, where oxygen in the air attacks any double bonds of an unsaturated fatty acid, releasing that energy.[39] This does happen as side reaction, but is not fundamental to Turkey red. A full explanation of how the cotton attracted oil was elusive because the analysis must be done *in situ*, or on the fiber rather than on an extract. This was not possible until more advanced techniques were available in the twentieth century. Research in the 1960s analysed historical Turkey red using infrared spectroscopy, and found oily material on the cotton but did not identify how it was bonded to the fiber. Practical testing found a degree of hydrophobicity, which the authors attributed to the oil, which they proposed may also influence the deposition of the color complex.[40] This is consistent with earlier observations of the cut fibers having white cores, and supports the theory of its physical function providing some protection from water via the hydrophobic fatty acid chains. Recent research has finally answered the chemistry of how the oil treatment becomes part of Turkey red. *In situ* infrared spectroscopic analysis of oiled cotton samples and Turkey red showed the carboxylic acid end of the fatty acids in the oil treatment (see Figure 3.3) forms hydrogen bonds with hydroxyl groups on the cellulose rings (see Figure 3.1). This analysis also functions as a non-invasive screening method to identify Turkey red based on whether signals from fatty acids are present. It should be noted that similar results may be seen with alizarin and madder reds as well, but there is no comparative study of these with Turkey red to date. Hydrogen bonds are relatively weak compared to other bonds, so they break more easily, explaining why excessive degreasing can strip the treatment from the fibers.[41] The fiber remains stable because its structure is not altered by the fatty acids attaching to it. Hydrogen bonds were not known until the early twentieth century, so although Turkey red dyeing had been developed, refined, and researched extensively, the knowledge to explain it was not yet available.

It is unknown how the oil treatment was originally developed. Perhaps it was discovered accidentally, since its chemistry is effectively an oily soap solution slow-dried onto cotton. The same deposition of fatty acids could occur over repeated washings when cotton is laundered with traditional soap, which is chemically different from the laundry detergents used today. If this cotton were later dyed (perhaps something that was dingy, but still useful), the resulting color would no doubt provoke further experimentation to replicate it.

Ruminant Dung and Tannins

For today's reader, one of the more notorious aspects of Turkey red dyeing is the addition of animal dung to the oil bath, where it functioned as an assistant. This is primarily associated with the old process, though archive records from the late nineteenth and early twentieth centuries show continued use of dung well after the new process was dominant.[42] It was common practice to have two stages of oil treatment—the white bath already discussed, and the "gray" or "green" bath (*bain de fiente* or *bain bis* in French). It was prepared by trampling the dung of ruminant animals like sheep and cows into a white bath, a task sometimes assigned to young boys. In some early methods, "intestinal liquor" is used instead.[43] Methods with the gray bath typically direct three applications of the bath interspersed with drying, washing , then three applications of the white bath with drying, before washing and drying again.[44] The purpose of dung was often explained as "animalizing" the cotton, a reference to the relative ease of obtaining deep colors on wool and silk.[45] Some said the dung was not essential, and could be omitted without any effect to the color, while others believed it had a positive effect and was indispensable.[46] A late nineteenth-century source said sheep dung was not used in French dyeworks and was "peculiar to the Levant," but it appears in French and English dyeing treatises so the actual extent of practice is unclear.[47]

Another optional treatment that follows degreasing is a tannin bath, which also becomes less common with the adoption of the new process.[48] Tannins are a broad class of natural compounds sourced from many plants, familiar to us for the astringency they give to tea, unripe fruit, and red wine. In India and Indonesia, dried myrobalan fruit (*Terminalia chebula*, *Terminalia bellirica*) was commonly used.[49] In Europe and the Eastern Mediterranean, tannins for dyeing often came from oak galls (also called gall nuts) and sumac plants, so the step was sometimes called galling (*engallage*) or sumaching. Gall nuts form when female wasps from the genus *Cynips* lay eggs in the new branches of some species of oak trees, which respond by forming a swelling around the larvae which contains a significant amount of tannic acid. The best ones, often called Aleppo galls, came from the Eastern Mediterranean and contained about 60 percent tannic acid by weight. Sumac (sometimes spelled sumach), from the leaves and twigs of several species (*Rhus coriaria*, *Rhus cotinus*, and *Coriaria myrtifolia*), contained less tannic acid (about 20 percent acid by weight) and the best quality was found in Sicily. The higher tannic acid content made galls the better option, but they were also more expensive so some dyers used a mixture of galls and sumac.[50] A tannin bath was prepared by brewing the plant material in hot water, much like a tea. The degreased cotton was dipped or soaked in the hot tannin bath (around 50 °C/120 °F) for up to six hours, then pressed of excess liquid.[51]

The suspected purpose of the dung and the tannins was to facilitate the aluminium application. Using a solution of cow dung to cleanse fabric is an old practice, and appears in descriptions of textile dyeing and printing techniques from India, where it likely originated.[52] Nineteenth-century chemists experimented on the composition of animal droppings to understand their role in dyeing, and perhaps identify a chemical substitute. They found ruminant dung supplied mineral salts and organic compounds including phosphates, carbonates, silicates, sulfates, and chlorides of calcium, potassium, sodium, and ammonium. These complexed with excess aluminium to form a precipitate, preventing loose metal ions from complexing with the dye in the bath. Phosphates were particularly useful, since aluminium phosphate is insoluble in water, forming a precipitate. Since the desired components were water soluble, the dung should be fresh and not rained on.[53] These included pure forms of some compounds found in dung, like sodium phosphate. Sodium arsenate was found to be more effective than the phosphate since it was cheaper and its aluminium complex was less water soluble, however, arsenate salts had acknowledged drawbacks including damaging the workers' hands, contaminating waterways with runoff, and lingering on the fiber in trace amounts that had a chance of poisoning the wearer. To mitigate costs, some effort was made to recover dung substitutes after treatment for re-use.[54] Some said dung gave better results than the salt substitutes by acting more slowly and evenly.[55] Although its necessity in Turkey red was debated, dung or dung substitutes likely improved the efficiency of the dye bath and the rub fastness by complexing with poorly attached aluminium. Archive documents show the J&P Coats firm used silicates as a dung substitute in the early twentieth century, and cow dung was still used throughout by the Archibald Orr Ewing firm until at least the 1890s.[56] Although its value in Turkey red dyeing was debated, dung treatment was especially useful for prints since the dislocation of loose metal salts would spoil the design, which was determined by selective application of metal mordants.[57] Some printers kept as many as thirty head of cattle for a guaranteed supply, which was inconvenient and expensive.[58]

Cotton without an oil treatment was known to be poor at attracting aluminium ions. The oil pre-treatment was particular to Turkey red, but tannin treatments also improve the depth of color dyed on cotton and appear in many more methods. Tannins contain many hydroxyl groups, which also adhere to the cellulose via hydrogen bonding like the fatty acids. Other hydroxyls on the tannins form bonds with aluminium, iron, and tin mordants, acting as a fixing agent by forming metal salts on the fibers.[59] In Turkey red, tannins are not an essential part of the complex, but they do act as a supplemental treatment parallel to the oiling, effectively dyeing a madder red with the Turkey red.

Turkey Red Oil and the New Process

The rancid plant oils used in the old process contained a relatively low proportion of the desired free fatty acids compared to the undesired partial and full triglyceride oils. This generated a significant amount of extra work since multiple applications were needed to deposit enough fatty acid on the fibers, and remove the excess triglycerides by degreasing. Naturally, dyers and chemists sought a means to economize on this step to reduce time, labor, and materials used. The goal was to find an oil with a high concentration of water-soluble fatty acids, which would make the treatment more efficient in both steps. This was accomplished by chemically treating plant oils to break the triglyceride bonds, effectively a more complete hydrolysis compared to natural rancidification. Experiments on treating olive oil with concentrated sulfuric acid were published from the 1830s.[60] In 1834, German chemist Friedlieb Ferdinand Runge made a preparation of olive oil and sulfuric acid that he successfully used as a treatment to dye cotton red.[61] An 1804 Turkey red method includes sulfuric acid in the white bath , which indicates possible knowledge of its effect on the oil.[62] This was atypical practice, however, since highly acidic conditions would damage the cotton and acid-treated oil was neutralized before use.

By the mid-nineteenth century, there is evidence of prepared oil use and early commercial products. In 1847, Mercer and Greenwood patented a product called "sulphated oil," made by treating olive oil with sulfuric acid, though it was not widely adopted.[63] One Mulhouse dyer used olive oil treated with potassium chlorate around 1867, which supposedly shortened Turkey red dyeing to forty-eight hours.[64] Potassium chlorate is highly reactive and hazardous, so it was likely not worth the trouble which explains why the practice was uncommon. In general, acid-treated olive oil products were not very useful because the sulfuric acid reacted very strongly and caused "charring," where the oil darkened and could discolor the cotton.[65] Other options were sought, and many ruled out for their cost or reactivity, like linseed oil.[66] Castor oil, long used as a highly effective oral laxative, emerged as a strong candidate due to its chemical properties and acceptable price. It is unusual for being primarily the triglycerides of one fatty acid, ricinoleic acid, which comprises around 90 percent of the fatty acids present.[67] By comparison, olive oil is about 70 percent oleic acid. Ricinoleic acid (see Figure 3.7) has a very similar structure to oleic acid (see Figure 3.3), but with a hydroxyl group attached to the 12-carbon. These abundant hydroxyls in castor oil form hydrogen bonds with each other, reducing its flow and making it particularly viscous for an oil. These groups also react first with sulfuric acid, limiting the number of secondary and side reactions that occur with oleic acid, which make it less prone to charring than olive oil.[68]

Credit for the adoption of castor oil is given to both Horace Koechlin in Mulhouse and Walter Crum in Glasgow, who discovered its potential almost simultaneously around 1870.[69] Others give credit to a Dr. Wuth and Fritz Storck in the same year, so clearly multiple parties were investigating the matter.[70] The product of treating castor oil with sulfuric acid came to be known as Turkey red oil, or sometimes alizarin (or alizarine) oil, oleine, soluble oil, dyeing oil, and red oil. It was commercially available from the early 1870s, and quickly replaced rancid olive oil in many dyeing establishments.[71] Dyers could purchase Turkey red oil in concentrations from about 30 to 70 percent oil, the rest being water. More concentrated products reduced the overall cost of freight. Turkey red oil was sold by manufacturers like John M. Sumner & Co. in Manchester, England, P. L'Honore in Le Havre, France, and Mueller-Jacobs in Russia, but many textile firms made their own in-house. Archive documents from the J&P Coats firm describe an in-house method for making dyeing oil using a combination of olive and castor oil treated with sulfuric acid.[72] Its applications were never exclusively for Turkey red, and for many years synthetic dyes, including alizarin, were sold in a paste of Turkey red oil to facilitate transport and use. Even though Turkey red is no longer dyed, Turkey red oil is still used today in textile processing, ink making, industrial detergents, leather treatment, and as a lubricant additive and skin emollient. It was the first synthetic anionic surfactant, and led many subsequent developments for wetting and leveling agents in the early twentieth century.[73]

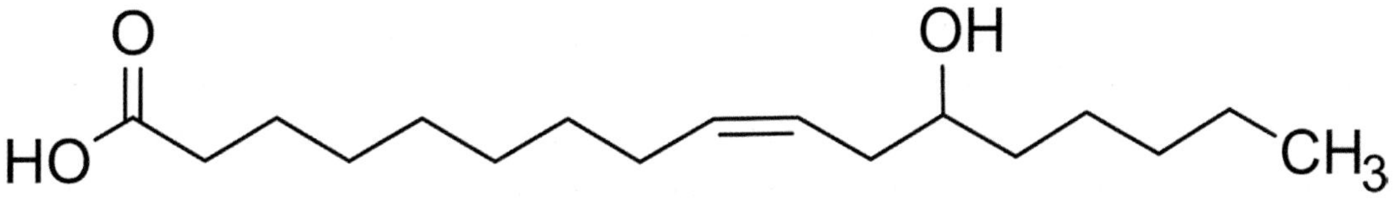

Figure 3.7 Structure of ricinoleic acid, the major fatty acid component of castor oil.

Making Turkey red oil is simple, and typical methods from the late nineteenth century are similar to industrial practice today.[74] Due to its hydrogen bonding behavior, castor oil has a similar consistency to honey. This means mixing it with thinner liquids like sulfuric acid requires care to ensure an even distribution. Historical methods recommend mixing the ingredients in a large tank, at least two to three times the original volume of oil, which will increase significantly over the course of the preparation. Typically, a 4:1 ratio by weight of castor oil and concentrated sulfuric acid (around 98 percent) was used. Sometimes oleum, the highly reactive mixture of sulfuric acid and sulfur trioxide, was used. The acid was added slowly with stirring to ensure a complete reaction, and to keep the temperature below 35 °C (95 °F) since it released a considerable amount of heat that could cause undesired side-products. The pale-yellow oil turns dark, cloudy brown and becomes less viscous as it reacts with the acid. The mixture was left overnight to react. The next day, it was stirred and added to a wash solution, either pure water or a solution of sodium sulfate, in a large tank with a tap at the bottom. This was also left overnight, during which the oil and water separate into two layers. The lower water layer, which contained most of the excess acid, was drained off and discarded.[75] Washed oil should be creamy and yellow, without any dark streaks. After washing, the oil was neutralized with a strong alkali like sodium hydroxide, potassium hydroxide, or ammonia.[76] During neutralization, the creamy appearance should dissipate. The final product should be a clear, golden yellow without any cloudiness. Insufficiently neutralized oil will continue to react, causing turbidity and eventually ruining the oil as well as degrading the cotton fibers during application.

The products of a reaction between a triglyceride oil and sulfuric acid are complicated and varied. First, the glyceride bonds are hydrolyzed, releasing the desired free fatty acids and glycerol, which is removed in the wash. Next, the acid reacts with any hydroxyl groups, then any double bonds. This is how ricinoleic acid is less reactive than oleic acid, which only has the double bond. Chemists researched the reactions and products of Turkey red oil, and investigated methods to test for quality and adulterants.[77] An 1887 thesis project at the Massachusetts Institute of Technology systematically tested methods to make Turkey red oil by Lightfoot, Mueller-Jacobs, Stein, Schatz, Lauber, Liechti and Suida, Schmid, and Lubianoff, followed by Turkey red dyeing tests.[78] Recent analysis compared the composition of commercial Turkey red oil, replica Turkey red oil made following historical documentation, and a sample from the late nineteenth century. The analysis showed the Turkey red oil samples had a significantly higher acid content, corresponding the desired free fatty acids, when compared to unmodified castor oil. The free fatty acids have hydrophobic and hydrophilic ends, giving it good surfactant and wetting properties, which increases the liquid permeability of the fiber and facilitates dyeing. In hard water it may form insoluble, sticky calcium and magnesium salts, which is why adding calcium to the oil bath was not common practice. Turkey red oil is sometimes called sulfated castor oil, since sulfuric acid can react with the hydroxyl group on ricinoleic acid to form a sulfate ester. These compounds hydrolyze easily during neutralization, and the analysis found they are not a significant component of Turkey red oil. Overall, the new and historical samples had a similar, inhomogeneous composition consisting of free fatty acids and some partial glycerides from incomplete hydrolysis. The historical Turkey red oil showed evidence of the free fatty acids cross-linking into polymer compounds called estolides, which decrease the utility of the oil for dyeing since they are no longer available to bond with the cellulose. This process happens slowly at room temperature, so dye firms could avoid any losses by using and rotating Turkey red oil stocks.[79]

Oiling cotton in the new process was relatively simple. Turkey red oil is fully miscible with water and required no soda leys, only dilution. The treatment also did not need to be heated. It is composed primarily of free fatty acid sodium salts, which is also what forms in the white bath when the soda ley is mixed with rancid oils. To prepare 100 pounds of cotton for Turkey red, about 10–15 kilograms of Turkey red oil at 50 percent concentration was mixed with one hundred liters of water to make a solution around 5–10 percent concentration. This made a "slightly opalescent solution" that looks like thin, pale-yellow milk. The high concentration of free fatty acids meant only one application was necessary, and the absence of triglycerides reduced the need for degreasing. After saturating the cotton, it was pressed to drain the excess solution and then dried as usual.[80]

Sometimes the oiled, dried cotton was subjected to a slightly pressurized steam treatment (see Figure 3.8), about 2–5 psi for thirty to sixty minutes, which was thought to help the oil penetrate further into the fibers. Excess oil still had to be removed before the aluminium treatment or it would contaminate the bath.[81] Steaming probably removed some, but with Turkey red oil degreasing was easily done by washing once in clear water.[82] After oiling and washing, the dyer proceeded to the aluminium treatment, which did not change between the old and new processes.

Félix Driessen, the Dutch dyer and chemist, tested samples of Turkey red from other manufacturers and found differences in quality between old and new process samples, though he determined that both treatments fulfilled the same purpose.[83] The old process samples generally had better fastness, but this was not universally true and Driessen thought the difference could be overcome with systematic testing of the new process.[84] Indeed, dyers did not universally adopt the new process and many Scottish firms stayed with the old process, having the opinion that it made better Turkey red.[85] Archive records from the Archibald Orr Ewing firm show Turkey red oil was not followed until the mid-1880s, and olive oil was still partially used as of 1892. Early twentieth-century documents from J&P Coats mention the use of "Gantert's Tournante Oil," indicating olive

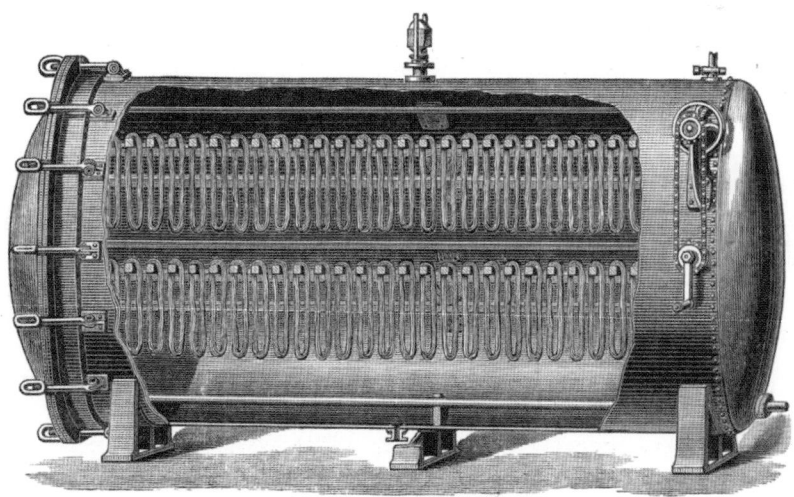

Fig. 94.—Steaming-Chest for Turkey-red Yarn.

Figure 3.8 Pressurized steam chamber for finishing Turkey red yarn. *The Dyeing of Textile Fabrics. Hummel, John James. Cassell & Co., London. 1885*

oil was also used for some time after the new process was available.[86] The Dutch firm Kaulen and Carp also continued to use the old process into the early twentieth century.[87] It is not yet known whether there is any material basis for the old process making a superior product, or if it is possible to distinguish oiling treatments analytically.

The Steiner Process

A third method for applying the oil treatment in Turkey red was developed in the nineteenth century, though it was not widely adopted and appears to be primarily, if not exclusively practiced by its inventor, Frederick (or Frédéric) Steiner. Steiner was an Alsatian dyer who studied with Daniel Koechlin from 1810 to 1811, around the time Koechlin devised his means of dyeing Turkey red cloth. In 1817, he immigrated to Lancashire and worked for the Broad Oak print works near Accrington, north of Manchester. Broad Oak was one of the most prestigious workshops in the area, and Steiner was named head of the chemistry laboratory. In 1836, he opened his own Turkey red establishment near Manchester.[88]

The Steiner process involved heating "clear" olive oil (not rancid) to 110 °C (230 °F), saturating the cotton in the hot oil, and pressing it to remove excess before hanging in hot stoves around 70 °C (160 °F) for two hours. Unlike slow drying in the old process, the temperature should be raised as quickly as possible. Afterward, the oiled cloth was soaked in sodium carbonate solution and dried in the stove, which was repeated a few times. The cotton was degreased and washed to finish the treatment before proceeding as usual with the aluminium.[89] Steiner's method was also adopted by his nephew, Charles-Émile, who worked with his uncle in England for some years before returning to France upon his father's death. The younger Steiner implemented the method at the family firm in Ribeauvillé, near Colmar in France. In recognition of his knowledge and expertise in the industry, he was invited to become a member of the prestigious Société Industrielle de Mulhouse in 1842.[90] Unfortunately, little primary documentation remains since the archives of Frederick Steiner were destroyed in a fire in 1950.[91]

An unusual example of Turkey red velvet dyed by the Steiner process is included in the 1846 *Traité théorique et pratique de l'impression des tissus*.[92] It is not clear whether the Steiner oiling was more suitable for less structurally robust pile-woven textiles like this, or if it was merely a specialty product of the firm. There is very little evidence or discussion of Turkey red velvet in general. Available literature and documents do not indicate any other large dye establishments using the Steiner process, so it is difficult to say whether it was unpopular or merely unknown. Chemically, the effect of heat and alkali treatments would create the same fatty acid deposit on the cotton fibers, but without more research the advantages of this method are unclear.

Aluminium

The oiled cotton was ready to be treated with aluminium, a step independent of the oiling method followed. As discussed in Chapter 2, dyers in India and Indonesia relied more on botanical sources for aluminium, like *Symplocos* and *Memecylon*, whereas dyers in the Eastern Mediterranean and

Europe used mineral alum, or potassium aluminium sulfate.[93] Some texts specify Roman alum, which was mined from pits in Tolfa, near Rome, that were opened in the fifteenth century by Pope Pius II.[94] It was considered to be the purest available for its very low iron content, and dyed brighter, clearer colors.[95] British alum from Lancashire, Yorkshire, and southern Scotland was inferior to "roach" (from *roche*, or rock) alum from İzmir and Roman alum.[96]

The bath was prepared by dissolving alum in warm water, then slowly adding a soda solution to make it alkaline. They react to form aluminium hydroxide in solution, but an excess of alkali will precipitate the aluminium which makes it useless for dyeing. Each addition of soda causes bubbles to release and white flocculation to form and dissolve as the aluminium reacts. When the white flocculation persists, the reaction is complete.[97] Another option was a solution of aluminium acetate, often made by mixing aqueous solutions of lead acetate (sometimes called "sugar of lead" for its sweet taste) and alum. The compounds exchange ions in solution to precipitate lead sulfate as a white solid, leaving dissolved aluminium acetate. It was often called "red liquor" because of its regular use in madder dyeing and printing.[98] Red liquor could also be made by dissolving aluminium hydroxide in acetic acid, then diluting with water.[99] To treat the cotton, the aluminium bath was warmed to around 40–50 °C (100–120 °F). The cotton was worked until saturated, then left for a period of about four to twenty-four hours. After wringing, it was hung to dry, washed in clear water to remove any excess material, and then dried again.[100]

Some said red liquor was not advantageous and cost more since it required additional reagents, while others argued that red liquor dyed better colors than alum alone.[101] Both appear in dyeing treatises, and choice was probably based on personal experience, cost, and the availability of ingredients. Although iron was a contaminant in red dyeing, if a controlled amount of iron sulfate was added to the alum it dyed durable violets and purples, effectively Turkey purple.[102] Maison Weber in Mulhouse was particularly known for their "*violets façon rouge turc*," but in general there is very little evidence of Turkey purple production. It may have been less popular, or Turkey red firms may not have wanted to take the risk since trace iron contamination could ruin an entire batch, and as such it was often made in a dedicated establishment to avoid this risk.[103]

Precipitated Aluminium Soaps

Nineteenth-century chemists knew that introducing aluminium ions to the fatty acids on the cotton was likely to form a kind of aluminium soap. Metal soaps like this are made from heavier metals like calcium, copper, aluminium, lead, etc., instead of sodium like in common soap, and are mostly insoluble in water. In the aluminium bath, the fatty acids from the oil treatment are converted to compounds like aluminium oleate, which was found to have an impressive affinity for binding some colorants. Like the chemistry of the oil treatment, it was difficult to prove aluminium soaps were precipitating without better analytical techniques. Research in the 1960s confirmed the presence of aluminium deposits on historical Turkey red. More recently, infrared spectroscopy of Turkey red replica and historical samples identified aluminium soaps, confirming the fatty acids hydrogen-bonded to the cellulose are able to complex with an aluminium ion. This chemistry is also seen in oil paintings, which release fatty acids as they age that form aluminium soap complexes with certain pigments.[104]

The formation of aluminium soaps on the fibers through the application of oil, then aluminium, is a key aspect of the fastness of Turkey red. The hydrogen bonds of the initial oil treatment are stable enough to withstand moderate washing, but the low water solubility of aluminium soaps make the fatty acid layer even more durable. The reason oiling and aluminium are distinct steps in a true Turkey red process, and why sufficient degreasing is essential before proceeding, is to avoid precipitating aluminium soaps in solution instead of onto the fiber.

Dyeing

Despite the number of preparatory treatments, cotton ready to be dyed Turkey red is still white (off-white if tannins are applied) since the oil and aluminium treatments impart no color. The dye bath was prepared by filling the chosen vessel with cold water and adding the dye (either madder or synthetic alizarin paste) in a quantity determined by the original dry weight of the cotton. The prepared cotton was immersed in the cold bath and the temperature raised to boiling over an hour, then maintained for another thirty to sixty minutes or until the color was satisfactory.[105] Some boiled for much longer, while others recommended maintaining heat at a lower temperature around 70 °C (160 °F) for optimal coloration.[106] At the end of this seemingly unremarkable step, the once-white cotton emerged a brilliant, vivid red. It was rinsed in cool water, then hung to dry. Altogether, the amount of effort and time required to carry out the actual dyeing step is negligible compared to the extensive prior preparations, but the success of the color was entirely dependent on them.

Calcium is an essential ingredient in the dye bath, though for some time its role was also unknown. Its significance likely contributed to some of the mystery surrounding the process, and the difficulties establishing it in new locations. Initially, calcium was probably unknowingly in the water supply. Although hard water was usually undesirable for dyeing because the mineral content may interfere with the ingredients, small amounts of dissolved mineral assisted in some recipes.[107] Dyers in India in the mid-eighteenth century used a certain kind of hard water that they knew improved reds dyed with chay, their dye source.[108] Calcium could also be introduced via the madder. As discussed in Chapter 2, calciferous soil improves the quality of the roots, which absorb calcium that becomes part of the dye bath. In these cases, calcium was present but not added as a separate ingredient, meaning the dyer may have been unaware of its presence and therefore its significance.

Ultimately, although calcium can be supplied through moderately hard water, a dyer has better control over the process using pure water and adding a controlled amount of calcium. Published methods usually include an amount of added chalk (calcium carbonate) in the dye bath. Credit for this goes to Jean-Michel Haussmann, a dyer and apothecary working in the late 1770s in Logelbach, a town in France near Mulhouse. He had successfully dyed Turkey red in Rouen, and found his results after relocating were unsatisfactory. Haussmann determined through systematic testing that the water in Rouen contained dissolved calcium carbonate, while the water in Logelbach was nearly pure, and that adding chalk to the dye bath resolved his problem.[109] In his articles on the topic, published in 1802, Haussmann states that he had not found the ideal proportion but generally added an amount of chalk from one-sixth to one-fourth of the weight of madder.[110] Chalk was an acceptable calcium source, but another common mineral, gypsum (calcium sulfate), caused problems.[111] In one method, calcium is added to the oil bath rather than the dye bath. This is uncommon in European practice, but appears on the island of Pulau Flores, where Indonesian dyers add a type of coral (mostly calcium carbonate) to the oiling step, and on Borneo where they use a quantity of lime.[112]

Turkey red methods typically specify a quantity of madder about twice the original weight of the fibers (i.e., two kilograms madder per kilogram cotton).[113] With synthetic alizarin, dyers only needed about 150–200 grams of synthetic alizarin paste at 10 percent concentration per kilogram of cotton, a tenth of the madder weight. Entries from a ledger of Turkey red ingredients that begins in 1873 show synthetic alizarin was already in use, while madder was abandoned in 1874 and garancine in 1878. The firm switched to 20 percent alizarin paste in 1882.[114] The ledger entries list the average use per 100-pound (45 kg) batch over six-month periods, so it is not clear whether the colorants were mixed, although it is likely for shading purposes. The Dutch firm De Heyder began using synthetic alizarin in 1872 according to company records, which was supplied by Perkin since BASF was hampered by the Franco-Prussian War. Records from 1876 show trials comparing alizarin and garancine, which both cost roughly two guilders per batch (a current value of about € 27). The difference was not just in the expense, but also in an overall simplification of the dyeing process when synthetic alizarin was used. Dutch manufacturers, however, did not entirely abandon garancine as quickly as British ones did. This was partly because the shade it produced was slightly different from that made with synthetic alizarin, which was preferred in some markets, and probably because of the local madder industry.[115]

There is evidence that early synthetic alizarin was mixed with garancine for tinting, and the fact that commercial alizarin synthesis wasn't refined until the late 1870s indicates dyers were likely blending colorants in the dye bath. This implies a level of control over the final color despite the variable composition of the colorants, and raises the question of whether test batches were conducted for quality control. The shift to synthetic alizarin did not fundamentally change the dyeing step, but it was easier to disperse ten kilograms of alizarin paste than two hundred pounds of ground madder, not to mention disposing of the remaining waterlogged, woody root afterward. There were no madder particles to wash off the cotton, and dyeing with synthetic alizarin required less oiling and clearing. Once synthetic alizarin supply was reliable, it enabled dyeing more consistent colors than a natural product like madder, and significantly lowered costs.

Color Complexes in Turkey Red

A Turkey red dye bath of ground madder looks like a pot of loose tea brewing—dark, brownish, and murky. As the temperature rises to boiling, the color of the liquid shifts from a reddish brown to a deep, garnet red. The red color develops on the fibers as the aluminium attracts anthraquinones from the dye. This is often obscured by the madder particles, but lifting it out periodically reveals the formation of patchy pink deposits that develop into a solid, vivid red. When using synthetic alizarin, the bath is initially reddish-orange and semi-opaque, also developing into a deep, garnet red. Re-creating the process is almost magical, and one can really appreciate how knowledge of such a remarkable product came to be so coveted.

It was known that metals like aluminium facilitate the attachment of dyes to fibers, but the role of calcium was less clear. Experiments feeding cattle with the tops of madder plants, which were not useful for dyeing, caused them to produce reddish milk and yellower butter with no harmful effects. Sixteenth-century physicians observed that animals who consumed madder had also reddish bones. In 1736, English surgeon John Belichier published his treatise *An account of the bones of animals being changed to a red colour by aliment only*, which facilitated research on the growth and development of the skeleton. French polymath Henri-Louis Duhamel du Monceau determined through studies of animal bones stained by madder that it was the calcium binding with the colorant.[116] The approach was used for many years to study bone formation in young animals, but had the disadvantage that some animals, particularly cats, could not be induced to eat enough madder. It was later determined that pseudopurpurin is the colorant responsible for this staining.[117] Although the interaction was not characterized at the time, the evidence indicated a strong affinity between calcium and anthraquinone dyes.

Complexes of colored organic molecules and metal ions are often called lakes, or lake pigments, terminology typically used in the context of painting rather than textiles and dyeing. Lakes can be made with aluminium, iron, tin, and other metals. They have complicated structures that are difficult to characterize because they tend to be amorphous, or irregular, rather than crystalline. Historical literature mentions the "Turkey red lake," which refers to a complex of aluminium, alizarin (or others), and calcium. This can be made independently of the fiber by adding alum solution to a dyebath, which is why it was important to remove excess unbound aluminium beforehand. Experiments in the late nineteenth century determined aluminium-alizarin complexes were soluble in Turkey red oil. The mixture could be applied to textiles to color them, but it was not nearly as fast as Turkey red formed step-by-step following a traditional process. In this sense, the overall complex could be interpreted as a lake complex precipitated onto the fiber with an oil mordant—indeed, it has been called a "lake formed on vegetable fibre by combination of a fatty acid with calcium aluminium alizarinate." Textiles dyed without oil treatments are not considered to have lake complexes, however, and as mentioned the term is more commonly applied to paintings. Here, the term "complex" is preferred since it includes the oil as well and is independent of the usual associations for the term "lake."[118]

As with other questions about the chemistry of Turkey red, research was limited by the state of analysis at the time. Even today, there are not many analytical studies of metal-dye complexes, and few also consider the chemistry of the fiber. Alizarin-aluminium complexes have been investigated recently *in situ* on wool fibers, and *in vitro* (i.e., not dyed on fibers), but conclusions are still complicated by the amorphous structure.[119] Historical research on Turkey red investigated the ratio of alizarin, aluminium, and calcium, and found that the complex also contained water molecules. It was suspected that the hydroxyl groups on the alizarin bonded with the aluminium, and experiments studied complex formation under various conditions to observe the effect of functional group positions on dyeing capacity of individual anthraquinones.

There have been a few proposed structures for the Turkey red complex since the late nineteenth century, with little consistency on what elements of the process are accounted for. As an example, a proposed structure from the 1960s had two opposing alizarin molecules attached to one aluminium at the carbonyl and adjacent 1-hydroxyl, plus a water molecule, but did not consider the calcium or fiber. An attempt to obtain a more crystalline lake, which is more easily analysed, used an organic solvent, dimethyl formamide, to precipitate an alizarin-aluminium-calcium lake and the analogous purpurin lake.[120] This, however, is not chemically similar to an aqueous dyeing bath and since the solvent becomes part of the complex, these results are not representative of Turkey red. Analytical studies of the aluminium atoms provide more information and also considered the role of a cellulose analogue, though without any oil treatment. Unlike carbon, which is most stable with four bonds, aluminium forms multiple stable complexes with different numbers of bonds. One proposed structure has the aluminium bonded to two alizarins through the carbonyl and adjacent 1-hydroxyl, with some water molecules present at a ratio of 4:2:2 for alizarin-aluminium-calcium. The aluminium has a total of six bonds, and the calcium ions are sandwiched between opposing rings of alizarin molecules.[121] The structure looks like a double-winged butterfly with aluminium ions as the "body" and calcium ions suspended between the wing layers. A later project using computational chemistry determined that this proposed structure was the most likely to form based on bonding geometry and energy of the compounds present.[122] If the cotton enters the dye bath with aluminium soap precipitated on the cellulose, then it is theoretically possible for two alizarin molecules to complex with the aluminium and incorporate water molecules and calcium atoms. This may be a 2:1:1 ratio for alizarin-aluminium-calcium, or half the proposed complex, with the rest occupied by the fatty acid. This is a theoretical assumption that requires more robust analysis for confirmation.[123] Since the formation of Turkey red is dependent on the cellulose, fatty acid, aluminium, calcium, and dye being present, studies that omit any aspect will only provide a partial understanding of its structure. Nevertheless, the information can be useful when pieced together, perhaps in the future leading to an even a better understanding of Turkey red.[124]

Blood and Albumen

Another fabled aspect of Turkey red dyeing was the use of animal blood in the dye bath, typically from sheep or cattle, which supposedly gave more fire and brilliance to the color. Opinions varied on its value, and nineteenth-century chemists determined it had no role in the final complex. Its purpose was believed to be a complexing agent for impurities in the bath and that it removed the "brown and resinous matters" that affected the color. Like the use of dung in the oiling treatment, its role was initially attributed to an "animalizing" effect, implying it facilitated dyeing cotton the same hues as wool and silk. In Astrakhan, blood was mixed with the madder before adding it to the dye bath. Some said dyers in the East believed the blood possessed magical properties that assisted in fixing the color and contributed to the red shade, though no primary documentation has confirmed this. Scientific theory attributed its function to the albumen content, and that an alternative ingredient, egg albumen, did not produce the same results.[125]

The biggest problem with using blood was its rapid decomposition, so it was ideally used as quickly as possible. Slaughterhouses supplied blood for dyeing industry, which they collected in buckets and stirred with bundles of twigs to remove the fibrin. It was then emptied into a large tank of water, passing through a strainer. The clots collected on the sieve were broken down with a masher or mill and added to the tank. Dyers stored casks of blood in sheds or open-sided buildings far from other working areas as possible since the homogenized, diluted blood was often " in a state of commencing or advanced putrefaction, and stinking horribly." A doctor at the Health Office in Glasgow said the dye works were a problem for authorities due to the "effluvium nuisances proceeding from them." The smell emanated up to 140 meters and, unsurprisingly, caused nausea, loss of appetite, and diarrhea. Another problem was "exhausted clots" that had started rotting but had not yet been discarded, and it was recommended they were kept in closed casks and sealed when full. Ventilation systems were later introduced by building a chimney over a furnace and drawing the air through the fire to clear it.[126] About 130,000 gallons (roughly 600,000 liters) of bullocks' blood was used annually at the Dalquhurn works where dyeing was done by hand, "a most unhealthy and disagreeable occupation."[127]

Naturally, more convenient blood alternatives were sought, with various options developed during the nineteenth century. Blood albumen, sold as flakes, was extracted from clotted blood by separating it in tanks and drying it in pans. Pure albumen flakes are transparent or pale and without odor, a significant improvement over putrid blood. Albumen was also isolated from dung, but this was less pure.[128] Treatment with quicklime helped dry the blood, and was sometimes done in a kiln or stove.[129] Blood albumen appears in early twentieth century archive documents on Turkey red dyeing for the J&P Coats firm, decades after dyers abandoned madder for synthetic alizarin, which did not have the same impurities.[130] Similar to the dung and tannins, the use of blood or blood substitutes in the dye bath was beneficial to Turkey red, but not essential in the formation of the color complex.

Clearing

The finishing step for Turkey red was a "clearing" or "brightening" treatment (also called *avivage* or *rosage*). Clearing was essential, and without it the fastness and color would be suboptimal. As with the other steps, the details of clearing treatments varied but follow the same principles. The simplest involved boiling the dyed red cotton, which had been rinsed and dried, in an open vessel of soda ley for four to twenty-four hours. Adding bran to the clearing bath was sometimes practiced and supposedly improved the color. Other methods used a solution of ley and olive oil or tallow soap, similar to the white bath, which could also be reserved from the oiling step and used in clearing. Clearing was also done in a "closed steamer" (see Figure 3.9) for nine hours with leftover white bath from the oiling step. In terms of efficiency, a shorter clearing step facilitated production and reduced fuel consumption. Using a closed system instead of an open kettle raises the pressure and therefore the temperature, also reducing the need to replenish evaporated water over many hours of boiling. In the new process, clearing typically required less time and less soap due to the efficiency of Turkey red oil. Clearing for purple, puce, and black (made with increasing amounts of iron mordant) required some chlorine, added as bleaching powder. An effective clearing vigorously cleaned the fibers of any colorant that was not firmly adhered. The alkaline soap solution attracts any loosely adhered colorant or oil, functioning like a more aggressive degreasing. The effect of this treatment is quite obvious for madder-dyed Turkey red, which emerges dark red from the dye bath due from

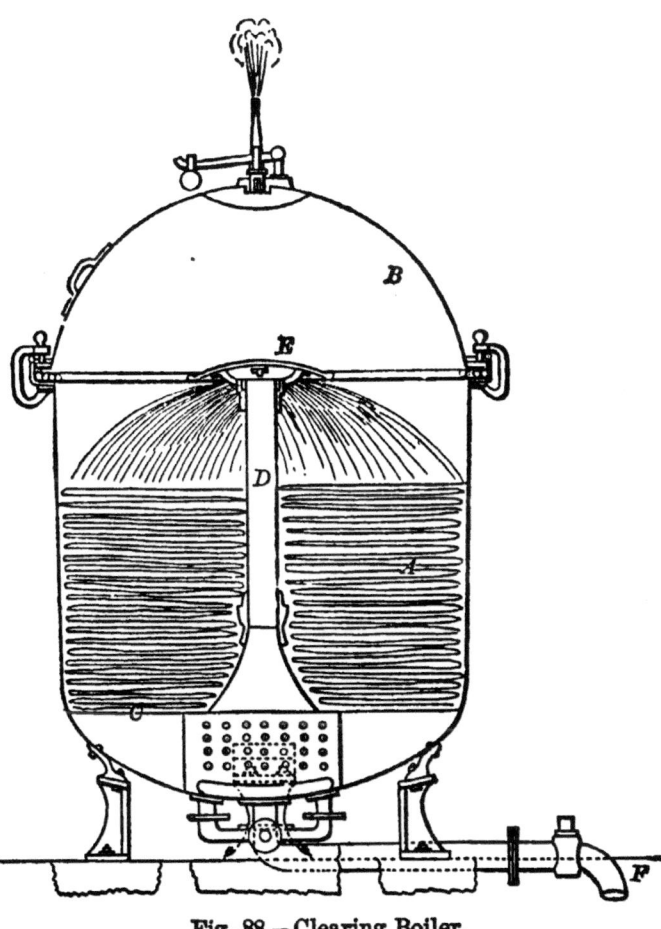

Fig. 88.—Clearing Boiler.

Figure 3.9 Diagram of a pressurized clearing boiler, late nineteenth century. *The Dyeing of Textile Fabrics. Hummel, John James. Cassell & Co., London. 1885*

excess colorant and undesirable root material, and becomes the characteristic bright, fiery red hue after clearing.[131]

Some methods include a small amount of tin salts or "tin crystals" (tin (II) chloride) in the clearing bath to brighten the color. The technique is said to have been developed by dyers Arvers and Saint-Evron in Rouen in 1785. In madder printing, tin salts were added to the aluminium mordant to counter the dulling effect of iron contamination. Tin mordants make brighter colors than aluminium, but they tend to make fibers sticky as a pre-dyeing mordant and must be used sparingly. Adding them to the clearing bath as a kind of finishing mordant made it possible to obtain some of the color while avoiding the stickiness issue. One theory was that the tin replaced some of the aluminium, shifting the tone of the color. Another was that the tin chloride altered some of the minor colorants, causing the color shift. Tin salts were also considered to be optional, and do not appear in all methods.[132]

After clearing, the Turkey red was washed once more and dried. Some establishments spread the finished cotton on fields of grass to expose it to sunlight. In Mulhouse, the was called "*La Mer Rouge*" or "The Red Sea" for the sight it created, and there is a neighborhood there today that bears this name.[133] Further textile finishing treatments like singeing, calendaring, and starching, may be applied, but are not particular to Turkey red.

Conclusion

The transition from the old to the new process significantly reduced production time and costs for Turkey red dyeing. Turkey red oil, the first synthetic anionic surfactant, and commercial synthetic alizarin, the first natural dye to be replicated, were both major technological advances for textile dyeing and industrial chemistry whose development was motivated by the demand for Turkey red. These materials were adopted by most dye firms during the 1870s, and although they revolutionized the Turkey red process, they did not fundamentally change its chemistry. The complex is built onto cellulose fibers through a series of treatments applied in a prescribed order. Fatty acids in the oil bath, from either rancid plant oils or Turkey red oil, form hydrogen bonds with hydroxyls on the cellulose. These become precipitated aluminium soaps in the subsequent aluminium bath. This was prepared with either mineral alum or parts of aluminium-accumulating plant species. The prepared cotton enters the dye bath, where the aluminium ions attract anthraquinone dyes. These complex with calcium ions, which were either present in the water supply or added, usually as chalk. Dyeing was primarily done with madder root, and also chay root in India, before the adoption of synthetic alizarin. Research found novel anthraquinone dyes anthrapurpurin and flavopurpurin in synthetic alizarin, which create slightly different shades of red and have a different chemical profile from madder. Many methods include the use of assistants like dung and dung substitutes, tannins, and blood or blood albumen. They are not necessary to make Turkey red, but they can improve the quality of the product. The chemistry of Turkey red shows why processes with regional and technological variations in ingredients can all be considered authentic since they create the same product on a material level.

4

Printed Turkey Red

Chapter Outline

The craze for printed cottons imported from India that began in the seventeenth century led to a nascent European textile printing industry. These textiles are sometimes called calicoes, a name which comes from the city on the southwest coast of India formerly called Calicut (now Kozhikode). As restrictions in France and England on cotton textile manufacturing were lifted, the industry expanded rapidly from the late eighteenth century, fueled by improvements in engineering, mechanization, and chemistry. Cotton printing was of sufficient interest to the general public that an article describing the processes and technology of block, plate, and cylinder printing appeared in an 1852 issue of *Godey's Lady's Book*, whose readers were the target market for many of these textiles.[1] Turkey red prints were a major textile product in the nineteenth century, and about half of the Turkey red dyed in Scotland was then printed.[2]

Printed Turkey red relies on some clever chemistry. Typically, printed textiles are made by layering colors over a white ground, but it is impossible to print Turkey red directly since the oil treatment cannot be selectively applied with reliable results. This indicates that although *kalamkari* appears to be a printed application of a Turkey red process, it was not considered such. Its deep red hue limited what colors could be printed over it, and for a long time black was the only viable option. This changed in the early nineteenth century with the development of discharge printing, which enabled the selective removal of Turkey red to create white areas that could be filled with other colors. In theory, any color that could be printed was eligible, but the hue and fastness of Turkey red created some limitations. First, the brilliant red does not blend visually with softer shades like pastels or buffs, so contemporary taste in textile design eliminated many options. The exceptional fastness of Turkey red also meant added colors needed to be similarly robust, or else

Figure 4.1 An 1857 Turkey red print showing the typical colorway of green, yellow, blue, black, and white on red ground. *University of Glasgow Archives and Special Collections, Records of United Turkey Red Co Ltd. GB248 UGD 13/8/4*

the textile would undergo differential fading. The two main colorants printed on discharged Turkey red were pigments, lead chromate and Prussian blue, which have low water solubility and good light stability. Logwood, a dye, was sometimes used for making black, but it is less fast and these prints are as stable. The bright, characteristic palette of most printed Turkey red makes visual identification easier than for the solid red fabric, though as always without analysis the assumption should be acknowledged as such. The most recognizable Turkey red prints like the one shown in Figure 4.1 contain a palette of red, yellow, green, blue, white, and black. This chapter discusses some relevant printing techniques, the development of discharge printing on Turkey red, and the process and chemistry of the colors that make its distinctive palette.

Textile Printing Methods

Printing involves transferring a design onto the desired surface via an apparatus. Designs can also be applied with a brush or stylus; these textiles were often called "painted." For centuries, textiles were printed using carved wooden blocks onto which the color (or often, mordant) was applied before stamping the fabric. In the dye bath, the mordanted areas selectively attract color to form the pattern, while assistants like dung complexed with any loose mordant that might otherwise cause staining or bleed. Polychrome designs are built using a system of pins or a grid to align or "register" the pattern since each color is applied separately. A major development was the addition of copper strips or pins to wood print blocks, which were more robust than wood alone (see Figure 4.2).[3] In block printing, printers worked in pairs on opposite sides of a table. The cloth was fed from beneath and drawn down the other end onto a drying cylinder. Color paste was applied to large blocks by a boy, called a tearer, who coated a woolen cloth stretched over a frame set up much like an inkpad. The printer pressed his block on the pad, the tearer re-wet the wool, and the printer registered the block by lining up brass pins before striking it with a hammer (see Figure 4.3). Separate pastes were applied for multiple colors, each of which had its own block corresponding to that part of the pattern. Two printers could work simultaneously, each doing one color.[4]

Attempts to partially automate printing by using rollers started in the eighteenth century. The most successful was in 1783 by a Scottish textile worker named Thomas Bell, who is usually credited as the inventor of printing with engraved cylinders, while working for the Hargreaves firm in Lancashire. By 1785, Bell had developed a six-color printing machine that matched forty block printers working by hand. He was likely also the first to use a steel blade, called a "doctor," positioned to scrape excess colorant from rollers between application and transfer, which made the process even more efficient. Cylinder prints were faster for uniform designs, and could have borders running along the selvedge, but designs that had a border on all four edges were still done by block. Hand block printing produced about 170 yards (150 meters) of fabric daily, whereas machine printing was capable of 5,600 to 14,000 yards (5,100–12,800 meters) daily. Copper plate printing

Figure 4.2 Wooden printing block with design outlined in copper, 1825–50. *Printing Block (USA); wood, copper, brass. Cooper Hewitt Smithsonian Museum of Design. 1941-87-1-h*

Figure 4.3 Engraving of a block printer and tearer. *La France industrielle ou Description des industries françaises. Poiré, Paul. Paris : Hachette, 1873. Source gallica.bnf.fr / BnF*

was also used, but was rapidly replaced in most instances by the arrival of more efficient cylinder printing. London printers, who famously used fine copper-plates, strongly criticized the introduction of machinery in the trade by the Lancashire printers as degradation of the art. Although block printing was not entirely abandoned, the proliferation of printed cottons in the nineteenth century, including Turkey red, would not have been possible without mechanical cylinder printing. Few of these cylinders survive in collections since they were large, heavy, inconvenient to store, and still had value as scrap after a pattern had been retired.[5]

Discharge Printing

Discharge (*enlevage* in French) printing was the only means of obtaining patterns on Turkey red. Under alkaline conditions it resists chlorine bleaching, but with a weak acid like citric or tartaric acid the color vanishes quickly.[6] There were two methods of application. In one, the pattern was created using lead plates carved with the desired motif. These were clamped onto the fabric at high pressure with a hydraulic press to create a resist, and the cloth was exposed to an acidified solution of bleaching powder (calcium hypochlorite). The areas not protected by the plates were discharged to white, while the rest remained red after the bleach was washed away. The second technique involved using blocks or rollers to print an acidic paste over all areas to be made white, and exposing it to calcium hypochlorite solution. This reacted with the acidified areas to remove color selectively, though it could not be left too long or the dissolving acid would affect the red areas as well.[7] The release of chlorine gas, which harmed workers and required additional ventilation, was a persistent issue.[8]

Lead Plate Press Discharging

The plates used in hydraulic press discharging had to resist corrosion from the highly caustic, acidified bleach solution, and lead was resistant to its effects. Designs were carved into the plates to form cavities, which corresponded to white areas. When clamped together and filled from above, the bleaching solution filled the cavities while the raised areas sealed off the cloth that was to remain red.[9] The most famous early use of this process was the Glasgow bandana factory of Henry Monteith, which employed the method to make red and white handkerchiefs. An 1824 account of a visit to Monteith's by pioneering Scottish chemist Andrew Ure in the *Glasgow Mechanics' Magazine* provides detailed mechanical descriptions of the "hydrostatic press," the plates, and factory operations. Workers stretched a stack of twelve pieces of Turkey red cloth and fed it between the printing plates for clamping and discharge. Monteith's factory operated sixteen presses simultaneously, with each stack taking about ten minutes to discharge. Four workers produced 19,200 yards (about 17,500 meters) of bandanas in ten hours.[10]

According to the account, the system was implemented there in 1818 by site manager George Roger. Attribution to Roger in Ure's account provoked a dispute between multiple parties in the local industry, carried out through letters to the editor published in the same magazine. It begins on February 21, 1824 with a letter from the *Magazine* calling discharge printing "a source of greatly increased revenue ... of incalculable importance to this country." The authors give credit to the true inventor, John Miller (also spelled Millar), saying he successfully discharged Turkey red in 1802 and had earned no profit or credit from his work.[11] The next week, a letter to the editor from an anonymous author "H." says they were "not a little astonished" to see credit going to Miller, and that discharging Turkey red with a screw-press and copper plates was first accomplished around 1801 by Robert Tweedie, a Turkey red dyer at a factory in Blantyre, southeast of Glasgow. In turn, Tweedie had a dispute with a Colin McCallum about who invented the process, though both were deceased at the time of writing. H. writes that Roger discharged the first bandana *for sale* (emphasis added) in 1802 and still had the original mahogany plates. The editors of the *Magazine* indicate the author is believed to be Henry Monteith. They also include the text of an 1818 letter provided by Miller's family, which says that in 1802 Tweedie asked Miller to develop a commercial means of printing and then attempted to copy the apparatus without Miller's knowledge.[12]

On March 6, 1824, the editors of the *Magazine* published a notice that they had received another claim to inventing the process from a living person and "not from any ghost of the deceased claimants, (which, indeed, would have alarmed us)." The testimonies were reviewed.[13] The following week, the editors state that they are confident Miller invented a discharging process in November 1802 and did not have prior knowledge of another method.[14] This, however, did not settle the matter. Two weeks later, the editors write that they have received yet more claims. In the same issue, a letter from Alexander Harvey asserts that Roger should receive credit, and that he sold the first bandana in Paisley in July 1802, invalidating Miller's claim. His footnotes include testimonies that Roger bought Turkey red cloth from Monteith in June 1802 and could provide an invoice, and that two men named Robert Henderson and William Galbreath purchased his printed bandanas that July.[15] David Campbell also writes, stating that Roger was not familiar with the equipment for discharging Turkey red, Campbell himself is responsible for the process. The editors thank Campbell and Harvey for their letters, inviting Roger to provide documentation that he was able to manufacture "at quantity."[16]

In July 1824, Miller writes that regardless of what else was said, at the time of his invention he was unaware of any other process. He says that in December 1803, he showed his model press to employees of the Monteith factory, and this was unquestionable because he had a letter from Tweedie to prove it. The tone of Miller's letter is indignant and defensive, continuing to say that Roger's use of flat boards was ill-suited for large scale production and Miller's metal plate system was superior even though it required some refining since early tests contained too much acid and caused the press to cut the fabric around the design. Miller provides a pattern (see Figure 4.4), presumably made from the discharging plate, of "the first flower ever discharged on Turkey-red cloth" and a statement from a Monteith employee attesting that he saw multiple patterns printed by Miller in or before 1803. Miller closes by offering to show anyone his models at his home, and to leave samples at the publishers' office.[17] In November, Campbell responds with a claim that he was making discharge plates in 1801, providing a letter saying that Colin McCallum was indentured to make discharging liquid and work the presses from 1801 to 1808. He gives supporting statements from Blantyre employees that Campbell was using lead plates to discharge Turkey red as of 1801, when he was the manager of the works.[18]

The debate resumes the following summer when a July 1825 letter from Miller describes his early printing efforts in greater detail, dismissing the work of Campbell with the insult "[If] the company who employed Mr. Campbell and his assistants had sent them a common broom, with instructions to dip it in the common bleaching liquor, and shake it above the cloth, it would have produced better work than what they produced." He also calls Campbell ignorant for using tin in his plates and the other correspondents "mere boasters," providing a number of letters to support his claims.[19] Campbell's response in September calls Miller's effort "futile and inaccurate to a remarkable, indeed pitiable extent" and that Miller is childish and cannot claim invention of a screw-press, which has been around since the time of Archimedes. Campbell also provides various letters and employment histories to support his story.[20] Miller responds in January 1826 to refute various points in Campbell's last letter and demand he provide an Excise extract of duties paid prior to 1809 to prove his claims.[21] He returns in February with two more letters to support his own claims.[22]

The two-year epistolary dispute ends with credit hanging between Campbell and Miller; Roger himself does not seem to have written to assert any claims. The relatively recent discovery of chlorine bleaching compounds no doubt led to many experiments by industry professionals around the turn of the nineteenth century, so it is entirely possible more than one of these claims is legitimate. Roger's use of wood plates seems to be the least practical due to the high

Figure 4.4 Design of the "first flower ever discharged on Turkey red," early nineteenth century. *Statement relative to the Discharging Process of Turkey Red, by means of Presses. Miller, John. The Glasgow Mechanics' Magazine and Annals of Philosophy, 1824 (1) 29. 462–4*

pressure required, and Miller had to make some adaptations to avoid the plates making cut-outs in the fabric, but these are more engineering matters than chemical ones. It is also worth noting that both Blantyre and Barrowfield were owned by Monteith, and the internal connections within the larger firm could equally have facilitated someone hearing of an experimental process through a colleague and making their own improvements before claiming credit based on a technicality.

Knowledge of discharge printed Turkey red appears around the same time in France. Christophe-Philippe Oberkampf, founder of the famous Jouy-en-Josas printworks near Versailles, credited Scottish printer Robert Henry with introducing discharging at his factory between 1803 and 1806. Henry had been taken as a prisoner of war while touring French textile factories, and became a liaison between Glasgow manufacturers and Oberkampf after his release.[23] The date implies early dissemination of knowledge from Scotland, so it was likely multiple parties were aware of the method.

Acid Paste Discharging

The second method of discharge printing using acid paste with alkaline bleaching liquor was invented by Daniel Koechlin of Mulhouse around 1811. His contribution was a practical adaptation following the same chemistry used in the lead-plate resist method, which he was aware of and sought to imitate. His experiments included hot soap and hot cylinders, which produced uneven textiles and many burns. Accommodating for various difficulties with the available materials, Koechlin found a way to print a mixture of gum (the sticky, water-soluble plant excretion), pipe clay, and vinegar, then dip the fabric in bleaching solution, which removed the red underneath the acidified paste.[24]

In 1813, James Thomson, owner of the Primrose works near Clitheroe in Lancashire, was granted a British patent for Turkey red discharging using acid paste and bleaching solution. He lists many options, including citric, oxalic, tartaric, malic, benzoic, sulfuric, sulfurous, phosphoric, fluoric, boracic, nitric, muriatic, arsenic, tungstic, succinic, and carbonic acids, no doubt to expand his patent protection since most of them are never mentioned in any other text about discharging and would be quite excessive to use. Thomson mixed the acid with a starch thickener, and used it to stamp the design onto the fabric before letting it dry and then exposing it to bleach.[25] It is not clear whether this patent was enforced at any time, though it seems unlikely considering there are no accounts of disputes.

Recipes for discharge paste thickeners from the second half of the nineteenth century primarily contain materials that were already used in textile printing. British gum, or dextrin, was made from roasting starch to yield a water-soluble adhesive that could be removed by washing. Printers also used materials like gum tragacanth, a natural mixture of polysaccharides sourced from a few plant species in the Middle East.[26] Pipe clay was another choice, sometimes combined with another polysaccharide, gum Senegal.[27] There is not any consistent recommendation in the literature for thickeners, so choice was probably based on cost, availability, and preference.

A Bright Palette

After discharging, the white areas on the Turkey red cloth could be kept or filled with other colors. It should be noted that the record of surviving patterns may not be representative of actual practice since preservation is inconsistent. Although there is the characteristic palette of red, white, black, yellow, blue, and green seen in most printed Turkey red, other colors were possible and surviving samples show printers did not always limit themselves. Turkey pink grounds were also discharged and printed. Pink was often called "pale red," and when combined with true red the product was a "two-red" print. The combination was popular for upholstery fabrics, like the one in Figure 4.5.

The two-red technique was developed by the same John Miller of the lead-plate discharge process. Contemporary printers who examined his products suspected it was a partial discharging of the mordant. In fact, they were made by dyeing fabric Turkey pink, then printing a design in aluminium acetate mordant and dyeing again to develop the red areas. This technique already existed in textile printing, but had not previously been applied in Turkey red.[28] The aluminium acetate was sufficient enough to take the Turkey pink ground to a full red, probably due to the underlying oil treatment in

Figure 4.5 Upholstery fabric in "two reds" colorway made by dyeing Turkey pink and printing the red design in aluminium before dyeing again. *University of Glasgow Archives and Special Collections, Records of United Turkey Red Co Ltd. GB248 UGD 13/8/4*

the pink. Materially, the printed red areas are more like a madder red than a true Turkey red based on how the aluminium is applied, which may decrease fastness. Sometimes "steam" colors were applied with dyes like catechu or Persian berries, which have a lower fastness than Turkey red. Steam green over undischarged red gave a brown ground, called *fond bronze*. An unusual example of a Turkey pink with a pattern of teal, blue, and goldenrod is shown in Figure 4.6. It is not clear what materials were used on this print, but the palette is unusual relative to the rest of the collection samples. A multi-colored print could have little Turkey red remaining if it was not the dominant color in the design.[29] Figure 4.7 shows three strips of printed Turkey red with decreasing amounts of red left after discharging.

Black, Blue, Yellow, and Green

Koechlin made his first prints in black over Turkey red ground in a pattern of palmettes around 1810. He used Prussian blue, an iron-based pigment invented in the early eighteenth century that was later used in cyanotypes, laundry bluing, and traditional blueprints. Prussian blue is composed of iron in two oxidation states along with cyano groups. The standard name is iron(II,III) hexacyanoferrate(II,III), also called ferric ferrocyanide. The Roman numerals refer to the oxidation states of the iron atoms.

Figure 4.6 A sample of Turkey pink bearing a pattern of birds and flowers. The colorway of green, blue, goldenrod, and teal on pink is uncommon relative to the traditional palette. *University of Glasgow Archives and Special Collections, Records of United Turkey Red Co Ltd. GB248 UGD 13/8/4*

Figure 4.7 A set of geometric patterns showing varying degrees of discharge printing. The motif of small squares building a larger pattern is common and was probably designed for the Indian market. *University of Glasgow Archives and Special Collections, Records of United Turkey Red Co Ltd. GB248 UGD 13/8/3*

Although free cyanide is highly toxic, the groups in Prussian blue are tightly bound and do not pose a risk. In fact, Prussian blue is useful as an antiradiation treatment for cesium poisoning. Black was also made by printing iron mordant and dyeing with logwood (*Haematoxylum campechianum*), also called campeachy wood or *bois de campeche*. Logwood blacks were significantly less fast to light and washing than those made with Prussian blue, and did not receive any final washing or processing.[30] The technique of adding black over red was sometimes called *noir d'application*.[31]

Koechlin found that mixing Prussian blue with his acidic paste before discharging replaced the red with blue when the bleaching solution was applied.[32] The swatch in Figure 4.8 shows a Turkey red print with a blue ground and yellow flowers, with very little remaining red. Another means of printing blue was to mix indigo with potassium hydroxide and print over white.[33] This method appears in a late nineteenth-century account of calico printing at the Broad Oak works in Lancashire, where Turkey red was also produced. Indigo has good wash and light fastness, but dyeing and printing with indigo requires a chemical reduction-oxidation process that will not work with only potassium hydroxide. Patterns containing indigo were typically dyed by resist and direct application, or "pencil blue" required a stronger reducing agent like arsenic acid.[34] This practice is uncommon in technical publications and treatises about dyeing and printing, and there is little indication that it was used in Turkey red prints.

Figure 4.8 A print where most of the red has been discharged and replaced with yellow and blue, leaving only the red outlines of the flowers. *University of Glasgow Archives and Special Collections, Records of United Turkey Red Co Ltd. GB248 UGD 13/8/4*

Yellow was made by precipitating lead chromate onto the fibers. The pigment was known since the early nineteenth century, and its application on textiles was first explored by French chemist Jean Louis Lassaigne around 1820. Lassaigne found that by printing lead acetate onto the fabric and exposing it to aqueous potassium dichromate, yellow lead chromate precipitated onto the fabric. The technique was optimized for Turkey red by mixing the lead acetate into the acid paste, then discharging as usual to remove red while depositing the lead onto the fibers. Afterward, yellow was developed with the potassium dichromate solution, forming lead chromate in the printed areas. The print sample in Figure 4.9 is an 1820s French textile with red rosettes on ground that was mostly discharged and replaced with yellow. By adding an excess of calcium hydroxide to the potassium dichromate solution, the color could be converted to chrome orange (perhaps as seen in Figure 4.6). Another option in this color family were manganese bronzes and iron buffs, a light tan or khaki color, which could be applied in a similar way by adding mordant into the acid paste and developing the color with a wash. Colors that were made with a printing and developing method were sometimes called "raised" colors.[35]

Green was made by layering yellow and blue. The discharging paste contained Prussian blue and lead acetate, leaving blue with the lead base that was developed in potassium bichromate to precipitate the yellow over the blue and form green.[36] Chrome green (chromium (III) oxide) was also used as a raised color, though its prevalence in Turkey red is unknown and most texts describe green as a combination of lead chromate and Prussian blue.[37] A thin layer of Prussian blue could also be used to tint the red a darker hue. The sample in Figure 4.10 has an error in the top right quadrant where the Prussian blue used to darken the red ground is slightly off register, leaving a streak of blue across the white petals. In figures 4.11 and 4.12, a series of samples from *Traité théorique et pratique de l'impression des tissus* shows the process followed to create polychrome

Figure 4.9 An 1820s French print of a red motif on a yellow ground made by Charles Widmer, the nephew of Oberkampf. In this sample most of the red has been discharged and replaced with yellow. *Cooper Hewitt, Smithsonian Design Museum. Bequest of Elinor Merrell. 1995-50-29*

Figure 4.10 A printed floral pattern with large areas of discharged white and green replacing the red. The maroon ground was made by printing a light layer of Prussian blue over undischarged red so it was darker but not quite black. *University of Glasgow Archives and Special Collections, Records of United Turkey Red Co Ltd. GB248 UGD 13/8/4*

Turkey red prints. First, the acidified Prussian blue paste was applied to the red ground. The bleaching solution discharged red in the white areas and from beneath the blue. Yellow was added over the blue leaves to make green, and to fill some of the white areas. Finally, black (possibly logwood since it was not applied with the blue) was printed over the top to finish the design.

Rainbow prints, or *fondu*, were a French innovation that allowed variegation and shading. It was first developed for wallpaper by Michel Spoerlin and Jean Zuber fils in 1822, and adapted for cotton prints in 1824 by the Dollfus-Meig and Koechlin firms. They were printed by blocks stamped on sieves that were loaded with different colors through a special system, then spread by a roller to form shaded bands.[38] The technique was also applied to Turkey red prints, and a surviving sample of *fondu* blue, yellow, and green made by a Scottish firm is shown in Figure 4.13.

Identifying Colorants on Turkey Red Prints

To date, there is little technical research on printed textiles. For some colors, informed guesses can be made for colorants present or the technique used, but this is not a robust method of identification. Analytical investigations, however, can be limited by factors including cost and sampling

347. Fond rouge turc, avec impression bleu et blanc enle-
vage prêt à passer en cuve décolorante.

Nous ne reviendrons pas sur les observations que nous avons
déjà faites à l'occasion du blanc enlevage, § 625 , p. 237 ; mais
le fabricant ne doit pas perdre de vue que le bleu de Prusse est
attaqué par les alcalis, et, par conséquent, par la chaux, toutes
les fois qu'il y a un grand excès de cette base et que la tempé-
rature de la cuve est trop élevée. D'un autre côté, le chlore
ternit le bleu et le fait virer au vert ; il faut donc qu'il sache
proportionner la quantité d'acide qui entre dans cette couleur

348. Fond rouge turc, avec impressions bleu et blanc enle-
vage passé en cuve décolorante.

Figure 4.11 Page from *Traité théorique et pratique de l'impression des tissus* by Persoz,
1846. Sample 347 shows Prussian blue and white printed before bleaching, and sample
348 after having been bleached. *Courtesy of Harvard Library*

350. Même échantillon que l'éch. 349, dans lequel on a rentré un jaune d'application.

351. Même échantillon que l'éch. 350, dans lequel on a rentré en dernier lieu du noir d'application.

qu'il a touchées. Enfin, dans l'éch. 351, nous retroúvons de plus le noir d'application qui s'est conservé sur tous les points recouverts de jaune. Cette double rentrure opérée sur une toile garancée ordinaire ou même sur toile huilée, mais dans des dessins plus déliés, aurait certainement été moins régulière.

L'éch. 352, de la fabrique de M. Steiner de Ribeauvillé, a été exécuté de la même manière ; on a donc :

    1° Imprimé sur une toile huilée un blanc enlevage sur rouge ;

Figure 4.12 Sample 350 shows the addition of chrome yellow and sample 351 after black, probably made with logwood, has been added in the final step. *Courtesy of Harvard Library*

Figure 4.13 An 1857 fondu or rainbow print where the discharged area is filled with a gradient of green, blue, and yellow. *University of Glasgow Archives and Special Collections, Records of United Turkey Red Co Ltd. GB248 UGD 13/8/3*

requirements. *In situ* techniques are prioritized when applicable, and micro-sampling limits the impact on an object, but analysis is not possible in every case.

Infrared spectra of discharged white areas on Turkey red prints show that fatty material is still present.[39] Although the mechanism of the discharging process and its effect on the color has not yet been described, this indicates the reaction does not affect the oil as much as it affects the anthraquinone dyes. The cyano groups in Prussian blue respond well to infrared spectroscopy, producing a distinctive peak that allows for non-invasive identification in blue, green, black, and dark red areas of Turkey red prints. Elemental analysis by *in situ* X-ray fluorescence spectroscopy (XRF) may also detect Prussian blue by its iron content, which is not otherwise expected to be present since the use of iron was strongly cautioned against in Turkey red dyeing. There is not enough research to say whether the iron mordant in logwood black can be detected by this analysis, so detecting the presence or absence of Prussian blue by infrared spectroscopy is the most reliable method. XRF easily detects lead and chromium, which are present in yellow and green areas. An alternative technique for elemental analysis is scanning electron microscopy with energy dispersive X-ray spectroscopy (SEM-EDX). These instruments operate in a vacuum, requiring a small sample

Figure 4.14 Scanning electron microscope image (2500x) of yellow fiber from printed Turkey red showing particles of lead chromate on the cotton. *Analysis was performed by Peter Chung of the Imaging Spectroscopy and Analysis Centre (ISAAC), School of Geographical and Earth Sciences, University of Glasgow*

or an object (up to a few cm). The SEM can take high-magnification images, such as the individual particles of lead chromate precipitated onto a yellow fiber seen in Figure 4.14.[40]

Organic dyes are best identified by high-performance liquid chromatography (HPLC), which requires taking a fiber sample and extracting the colorant. This includes logwood colorants hematoxylin and hematin, which can be identified through comparison with known references. Madder and synthetic alizarin dyes are also analysed using this technique.

Design

In the early nineteenth century, changes in excise duty opened markets to more consumers and made printed cottons a staple of working-class clothing. The Industrial Revolution and mass production enabled the general population to access ornamental and decorative goods, formerly only within the purview of those who could afford to commission artisans. Pattern and color choice were paramount to successfully market printed textiles, and designs did not need to be particularly artistic to be popular. The British public were less interested in aesthetically designed, high-class hand block printed fabric than they were in the affordable, novel designs that could be bought cheaply and worn for a brief period until the next fashionable pattern was available—an early version of fast fashion.[41] From a manufacturing perspective, interest shifted toward using the best new production techniques and novel developments in dye chemistry to entice consumers.[42] This led to a new field, industrial design, where objects were made in stages like an assembly line. As a result, the design and manufacturing processes became separated, whereas previously both had been at the behest of one craftsman. Industrial production increased the number of hands involved in printing a new pattern, which flowed from the designer or pattern drawer to copyists, engravers, printers, and other intermediaries.[43] Records of painted paper designs, finished and in-process, are preserved in the National Museums Scotland collection from United Turkey Red (cat. no. A.1962.1266.10.5.4855 A and A.1962.1266.28.2.3622).

Industrial Design and Production

The mass production of designed objects invited illicit imitations since the goods were widely accessible, generating protests from those who had made the initial investments and hoped to profit from their work. The cost of implementing a new design could be significant in terms of intellectual property, labor, and materials. Legal protections varied with time and location, but their existence did not ensure exclusivity since the protection was often difficult to enforce. In England, the first legislation to protect textile design was introduced in the late eighteenth century following years of petitions and inquiries. Reputable firms complained that "base and mean copies" of their designs were sold by "parasite firms" who made cheaper and lower-quality imitations, undercutting the market and avoiding many of the up-front costs incurred when creating a new pattern. Design copying was often blatant, practiced sometimes even by those who complained of their own being stolen, and sometimes accomplished through industrial espionage.[44] Since manufacturers were known to copy designs and prints were made for broad marketability, it is difficult to identify the manufacturer of a printed textile based on design alone unless there is a corresponding swatch with the manufacturer details registered. This practice was the genesis of legal disputes around intellectual property and labor disputes to prevent workers from transferring between firms.[45] Designs were often made in-house, and also purchased externally from freelancers.

Southern English manufacturers lobbied for legal protection while those in the North were opposed because they relied more on copied designs, having less access to talented designers and knowledge of current fashion trends. Temporary measures were enacted in 1787, renewed in 1789, and made permanent in 1794. This did not have a significant impact on copying practice overall, though it did introduce some delays before imitations came to market. The Designs Registration Act of 1839 was implemented to offer protection to "ornamental" designs through the National Board of Trade as industrial production expanded. The register comprises a complex set of records with notes on ownership and "representations," a physical copy or sample of the design being registered. In the case of textiles this most often involved a fabric sample, though drawings, tracings, and photographs were also submitted. There were different design processes for dress and furnishing fabrics. Sometimes a pattern was used for both, but in general furnishing designs were more elaborate, took longer to produce, and were more expensive. They were more likely to be printed by hand blocks or copper plates, which were slower to apply. Furniture designs did not sell as rapidly, but had a longer shelf-life that could justify the higher investment. As always, piracy was a concern and undercut profitability. The disconnect between designers and the production process created difficulties. Copyists, who transferred designs to rollers, blocks, and plates, chose designs similar to popular ones already on the market to mitigate costs and avoided scaling the design as much as possible. Sometimes patterns had to be adjusted to accommodate repetitions or to fit on a cylinder, possibly requiring multiple re-draws.

Compared to Britain, the design industry in France was far advanced in terms of aesthetics and practice. Protection of designs in France began in the early eighteenth century and was more comprehensive than in Britain. Designers were regarded as a respectable professional class and could earn 8,000–10,000 francs annually, equivalent to £320 to £400 at the time and roughly £40,000 today. There were more freelance designers in Paris compared to London and Manchester, and many worked for the English market. This created an imbalance where British manufacturers spent a good deal of money buying patterns from continental designers and making them domestically where the manufacturing technology was more advanced. James Thomson of the Primrose works commented that the industry in France was "preeminent in the ornamental, that of England, unrivalled in the useful." The Government School of Design, the first state-supported art school, was established in Somerset House, London in 1837 to educate British designers for the benefit of British industry. A Glasgow branch was established in 1845 to support local standards and production. Despite these efforts, critics lamented the state of British design at the 1851 Great Exhibition at the Crystal Palace in London, which showcased new technologies and products of recent years. Surviving correspondence shows American textile printers commonly "borrowed" designs from European manufacturers, though the practice was not universal.[46]

Into the nineteenth century, printing blocks were made on-site by designers and engravers.[47] Hand blocks continued to be used for high-end designs, but cylinders could produce more economical ones at a much higher rate. At a Scottish Turkey red works, finished designs were sent to the factory where a zinc cutter prepared a separate plate for each color in the design. From there, a pantographer scaled the design to the actual size and engraved it on copper rollers. The etcher was responsible for varnishing the roller and exposing it to an acid bath, after which the engravers removed any irregularities by hand and checked the design before it was put into production. This branch of work was considered to be the most skilled in the industry.[48] Cylinder surfaces could also be re-used, unlike wood blocks, and mechanical means of engraving eventually lowered costs for cylinder production. It could take skilled craftsmen two or three months to engrave a copper plate or roller, making it important to exploit each to the fullest to obtain the greatest possible return on investment. With conventional designs, this was achieved by printing a range of different palettes, but with Turkey-reds the number of variations was more limited.[49] A more affordable means of engraving was invented, or perhaps popularized, by Joseph Lockett around 1808. Lockett was head of a Manchester engraving firm and his method involved softening and engraving a small cylindrical steel die, about three inches long and one inch in diameter. It was then hardened and tempered before being rotated under pressure against a soft steel cylinder to transfer the pattern in relief (design raised above the surface). This cylinder was then hardened and rolled against a copper roller to repeat the pattern as many times as needed, rather than hand carving the entire design. The system lowered the estimated cost of engraving a roller from £1 to £5 as of 1860 figures (£60 to £300 today).[50] One English manufacturer said the average cost of preparing a pattern was £11, of which £7 to £8 was for engraving, and that the design studio cost about £800 annually (£47,000 in the twenty-first century), employing two to three men and four to five apprentices with no expense for freelance purchases. Another said that out of two to three thousand designs his firm made per year, about 250 were engraved on rollers and 300 on woodblocks. The cost of furnishing designs was given as £10 to £35 in the mid-nineteenth century. By the mid-1870s, the struggle for market share pushed Scottish Turkey red firms to produce fewer innovative designs and multiple lines of "fancy textiles" and more standardized, traditional prints and plain red cloth.[51]

European Design for the Export Market

Early European cotton exports to the Ottoman Empire tended to imitate Indian products due to their established popularity.[52] In Mulhouse, Daniel Koechlin's earliest black-on-red prints were marketed as "merinos." These paisley patterns imitated cashmere shawls called "merinos," which were woven in Scotland as imitations of the more prestigious shawls made from the hair of Kashmir goats.[53] Swiss firms also adopted patterns imitating cashmere shawls, which bore a pattern of cone or seed-pod shapes that originated in Persia (now Iran), today commonly known as paisley.[54] By the mid-nineteenth century, the size of the global Turkey red market was in part the result of the care firms took to accommodate the wide range of preferences across cultures, often generating vastly different products. Company documents and correspondence show targeted market research and product design, particularly for British manufacturers selling in India. Products were also presented according to local custom. On the British market, textiles were sold by length from a long bolt and finished garments from racks or shelves with no wrapping. Indian consumers valued presentation, and dress pieces were often wrapped individually in decorative packaging or folded a particular way. An 1854 request from a consignment merchant asks that turban pieces be folded exactly fifteen by twenty inches in bundles of ten wrapped in paper. A June 1857 memorandum from Mumbai said that folded textiles should be tied with orange cord rather than black because the latter was "objectionable," and another order specified bundles of five pieces to be tied with "green silk tape." Prints were also designed specifically for pieces to be worn as saris and sarongs, with complex patterned ends and borders.[55] An example from 1880s is shown in Figure 4.15 and bears a pattern of dancing women, peacocks, and flowers with a border of flowers and peacocks, for the Hindu market in India.

The size of the Mumbai market made it the primary target in India for manufacturers. Commerce was facilitated by the many intersecting trading routes there, and there was effort to improve transport infrastructure with roads, railways, and shipping. Some of these advances were commemorated in printed Turkey red, like the design of a stylized steam train framed by a traditional peacock pattern from the 1860s (see National Museums Scotland cat. no. A.1962.1266.31.9.627).[56] Records from the William Stirling and Sons firm include a book entitled "Bombay patterns" which contains samples of cloth and letters from 1853–69 describing acceptable shades of green and other specific observations about what had market appeal. A painted design card from 1859 with images of peacocks, women milking cows, and men in local dress was sent to India and returned as "unsuitable," with no further explanation why. In 1868, John Matheson, a company director, traveled to India. His book, *England to Delhi: A Narrative of Indian Travel* describes the clothing worn by people he encountered. Matheson observed that Hindu and Islamic

Figure 4.15 Turkey red print depicting dancing women, flowers, and peacocks with a border edge, likely for the Hindu market in India, about 1886–8. *University of Glasgow Archives and Special Collections, Records of United Turkey Red Co Ltd. GB248 UGD 13/8/6*

women wore different patterns, but Turkey red was popular among both. Hindu customers would purchase images depicting dancing people while Muslims favored geometric patterns, and peacocks were popular among both.[57] Some markets were accustomed to the same hand-printed designs on their textiles appreciated for decades, and European printers could only succeed by imitating what they wanted rather than introducing something new.[58] Samples and designs were sent out for review, with assessments returned to the manufacturers who would then determine production. Consignments of textiles were also taken out to other cities and regions, but the variety of preferences and religious customs throughout the country made selling some difficult depending on the design and location. Local taste could be difficult for foreign agents to assess, and often appeared idiosyncratic. For example, a particular shipment of fabrics to Kolkata was flagged as being unsaleable for a period from March to April because they were unlucky months for marriages according to local priests. These seemingly vague fluctuations in preference made it essential to have informed local agents to continue producing appealing designs for sale.[59]

In the 1890s, to become more independent of the British Empire, the Indian government raised tariffs on imported goods.[60] Since India was a key market for Scottish Turkey red, manufacturers sought to expand their reach. One design included a lion holding a saber in front of a rising sun, emblems of the Qajar dynasty in Iran. Commemorative designs for the opening of the Suez Canal were for the Egyptian and Turkish markets, while others were of waves, bridges, samurai, dragons, and butterflies for an East Asian market. A print sample of a samurai on horseback is shown in Figure 4.16. One of the largest body of records on printed Turkey red is the surviving sample pattern books kept by firms, like those in the National Museums Scotland and other collections (see Figure 4.17). The presentation of some indicates they were intended for customer viewing, while others appear to be for internal purposes with cryptic annotations about cylinders and plates. Volumes were occasionally classified based on the printing technique (cylinder, plate, block) or purpose (upholstery, bandanas). Despite these records, the persistence of certain popular patterns and the rampant copying between manufacturers often makes it difficult to connect a print in an object to a manufacturer.[61]

Figure 4.16 Turkey red print bearing flowers and a samurai on horseback, possibly intended for an East Asian market. *University of Glasgow Archives and Special Collections, Records of United Turkey Red Co Ltd. GB248 UGD 13/8/6*

Figure 4.17 Pages from a Turkey red sample pattern book, late nineteenth century, containing a variety of printed designs. There are no notes regarding the intended market, and the irregular presentation indicates it was only for internal use. *University of Glasgow Archives and Special Collections, Records of United Turkey Red Co Ltd. GB248 UGD 13/8/6*

Printed Turkey red patterns include stripes, tartans, paisleys, flowers, birds, animals, butterflies, landscapes, portraits, commemorative scenes, humans in all sorts of poses, commemorative scenes, and more.[62] A description from the 1862 International Exhibition of printed Turkey red from a large Scottish firm said "angry lions scowl out upon the visitor, gorgeous peacocks spread wide their argus-eyed tails, war horses and their riders caper and curvette amidst peaceful camels, and the lordly elephant bears proudly his royal burden in the royal scene."[63]

Conclusion

Turkey red prints were a major product of the textile industry, and can often be identified visually from their bright palette and bold patterns. Since the oil treatment could not be applied selectively to make a real Turkey red, direct printing was not possible and fully-dyed cloth had to be selectively discharged using acidified bleach. The two methods used were lead-plate resist and applied acidified paste. Although textile printing is an ancient art, discharge printing was only developed in the early nineteenth century. Prussian blue and lead chromate, the pigments used to make blue, yellow, green, and sometimes black, were of comparable fastness to Turkey red relative to other printed colors that might result in differential fading. Prints were not restricted to these materials, however, as logwood was also used in black areas and prints contain other colors, but these pigments are the most common in the object record.

Pattern design was an expensive part of printed textile manufacturing and imitation was rampant. This makes it difficult to identify the manufacturer of a textile, though sometimes records like the Design Registration Act, which attempted to address the copying problem, can be informative. Turkey red printers designed and produced for targeted markets, making a diverse range of products to meet local expectations. The demand for cheap, fashionable textiles industrialized the design process, separating it further from production.

5

Turkey Red in the Industrial Revolution

Chapter Outline

The Industrial Revolution had a significant impact on textile production capacity, increasing output in places like Western Europe where the innovative technology was first developed. This affected global markets as neighboring countries like England and France sought to outpace each other and sell the most fashionable goods at high volume and low cost, something hand manufacturing could not compete with.[1] As discussed in Chapter 2, knowledge of Turkey red dyeing came to Europe through migrant dyers and industrial espionage. Early European production was in the Western Ottoman and Hapsburg territories and done by hand, with trade on a moderate scale. By the time Turkey red was established in France in the mid-eighteenth century, the domestic textile industry was also adopting the machinery that gave them a significant competitive edge over hand production. Early European Turkey red was not always as good as imported product, and the initial supply could not meet local demand.[2] The literature does not indicate whether different qualities of Turkey red were sold, or what exactly constituted a bad product—presumably poor or uneven color, crocking, and bleeding in the wash were immediate signs based on its reputation for brilliance and fastness.

The concept of advertising also developed during the Industrial Revolution as companies sought to access the purchasing power of a rapidly growing consumer base. Like many products, Turkey red sales were promoted through advertisements and brand association. For the Great Exhibition of 1851, the firm William Stirling and Sons made a handkerchief depicting the Crystal Palace and with other themed motifs. They were sold at a significant loss, but served to advertise for the firm at such a large event. Direct marketing to consumers in the form of swatch books designed for the public, like the "Sun and Lion Brand" booklet in Figure 5.1, were used by sellers to promote and

Figure 5.1 Sample book of Turkey red calico for Sun and Lion Brand, late nineteenth or early twentieth century. This company did not manufacture Turkey red but rather acted as a middle-man for retail and wholesale consumers. *Cooper Hewitt, Smithsonian Design Museum. Gift of Harvey Smith. 1967-20-35*

display their product. These, in contrast to the sample pattern books like the one in Figure 4.17, were finished to a much higher degree. An 1876 story published in *Scribner's Monthly*, an American magazine, is framed as a letter to the editor and reads like an advertisement for Turkey red. The author describes a hunt for red calico at the behest of his wife, who wanted to cover some furniture. During a series of shop visits, he is told "if you want that quality of goods in red you ought to get Turkey red" and that it looked just like the swatch from his wife but was "much better." He returns home and explains to his wife why it was the better product, to which his wife said "this Turkey red is a great deal prettier than what I had, and you've got so much of it that I needn't use the other at all. I wish I had thought of Turkey red before."[3] No doubt some of the magazine's readers were sitting at home on their own furniture, possibly now thinking of Turkey red themselves.

Scottish Turkey red is closely associated with "export tickets" or "bale labels," the trademark labels with which individual manufacturers packaged their goods. They came into use sometime in the mid-nineteenth century, and were used to indicate authenticity and product quality. The visual representations were intended to be associated with a particular manufacturer and were closely guarded through legal action, what would be considered brand reputation today. This was valuable when identifying intentional misrepresentation by fraudulent products, and when it was difficult to enforce copyrights. The labels also communicated with and targeted specific markets, especially where consumers and retailers did not speak English or were not literate. They were about the size of a postcard and comprised part of the final product presentation along with the wrapping. A textile worker said "[T]he big tickets of identification were printed with images of Hindu gods—a worker had to stack the right cloth with the right god—so—if their arms and legs were not counted correctly a person would be in trouble."[4] Surviving labels, often printed by lithograph, bear the names of manufacturers and of retailers, who received product consignments from the dye and print houses to sell on international markets.[5] Many examples can be found in museum and archive collections whose holdings include documentation from these firms. A late nineteenth-century label depicting a frog carrying a mouse for the Singapore and Penang market is shown in Figure 5.2, and a debossed gilt label with text in the local language is shown in Figure 5.3. The labels were also used as a means of market research when local agents returned those that sold well, allowing manufacturers to correlate the information to the product.[6]

Factory production created new working conditions and social systems as society shifted from rural to urban. Dyeing was hard work, requiring large volumes of heated water and manipulating heavy, wet textiles. Mechanization alleviated some of this work and created other hazards. The products of these industrial laborers were exported to countries where the processes had been initially developed, like India for Turkey red, undermining the local markets that once profited from hand production. This chapter discusses major Turkey red producing countries during the nineteenth century and the accompanying changes to working life and the global economy in the nineteenth century. The list is not exhaustive, and more Turkey red operations likely existed than are discussed here, but it provides a sense of the scale of the industry in the nineteenth century.

Figure 5.2 Printed bale label for Adams, Gilfillan and Co., a Turkey red retailer operating in Singapore and Penang on the Malay Peninsula, late nineteenth century. *University of Glasgow Archives and Special Collections, Records of United Turkey Red Co Ltd. GB248 UGD 13/7/4*

Figure 5.3 Gilt bale label from the Archibald Orr Ewing firm, late nineteenth century. *University of Glasgow Archives and Special Collections, Records of United Turkey Red Co Ltd. GB248 UGD 13/5/13/1e*

Turkey Red Industry by Country

Industrial Turkey red dyeing in Western Europe began during the third quarter of the eighteenth century. Early attempts in France and Britain had varying degrees of practical success, less so commercially. By the 1780s, however, accounts and publications show there were multiple stable Turkey red dyeing operations in both countries. As before, knowledge of the process disseminated through the expansion of dyeworks and the migration of skilled workers. By the end of the eighteenth century, Turkey red was an industry in Switzerland, France, England, Scotland, and become so by the mid-nineteenth century in the Netherlands and North America. There is also evidence of production in Germany, Russia, and Czechia, though it is unclear on what scale. Archive records and more general histories of textiles and trade in those languages likely contain more information yet to be discovered.

France

The French Turkey red industry was concentrated in Rouen and Mulhouse, two major centers of French cotton textile production.[7] Early French establishments, like the government-subsidized operation in Darnétal (see Figure 2.8), struggled with market appeal, capital costs, and the technical challenges of establishing dyeing in a new location. A successful industry (rather than individual dyeworks operating in precarious conditions) took until the 1770s to stabilize, probably in part due to the different water supply and wetter climate.[8] Jean-Michel Haussmann, who determined the need for calcium in the dye bath, had moved to Rouen from Colmar in the 1770s and was part of the early industry there.[9] By the early nineteenth century, an estimated 150 Turkey red dyers were operating in the area, employing approximately 1,500 workers between them and producing two million pounds (900,000 kg) of dyed cotton annually at a value of close to twelve million francs. The industry consumed about 1,000–1,200 cords (roughly 128,000 cubic feet or 3,620 cubic meters) of wood annually to operate the drying furnaces. One named firm operating in Rouen was Cordier & Braun, but there appears to be little other detail available.[10]

On the eastern side of France, printed textile manufacturing began in Mulhouse in the mid-eighteenth century, so local manufacturers were already experienced working with cotton textiles. To fulfill the growing demand for skilled labor, workers were recruited from nearby Switzerland and Germany. The revocation of the Edict of Nantes in the late seventeenth century caused many skilled Huguenot textile workers to leave France for these places, and in the intervening decades the skill gap created by their departure limited the advancement of French textile manufacturing. The government had also recently moved to protect domestic wool and silk at the expense of novel cotton printing. At the time, Mulhouse was an independent republic associated with the Swiss Confederation. After the outbreak of the French Revolution in 1789, changes in its customs status and an economic blockade affected the local cotton industry until Mulhouse joined France in 1798. Cotton dyeing and printing expanded during the Napoleonic era, adopting new technology to increase production.[11]

Just before the French Revolution, some Greek dyers in Rouen were let go from the establishments there and became itinerant dyers, taking the process to Alsace.[12] This may be how Turkey red dyeing began in Mulhouse, although if Haussmann came from Colmar (about 28 miles/45 km north of Mulhouse) knowing how to dye Turkey red, then knowledge was clearly already circulating. Daniel Koechlin came from a family of Mulhouse textile manufacturers. His paternal grandfather, Samuel Koechlin, began dyeing there in 1745.[13] Samuel had two partners, Jean-Henri Dollfus and Jean-Jacques Schmaltzer. Dollfus left the company in 1765 to start his own firm, which later became Dollfus-Mieg. The Schlumberger family also founded a cotton dyeing and printing firm in Alsace in the late eighteenth century. The Blackburn Museum collection has an 1837 dye notebook written by Monsieur Schwartz of the Schlumberger firm, perhaps a connection to the family of Ambelakia dyers in Greece. By the early nineteenth century, the Steiner firm was operating north of Mulhouse in Ribeauvillé, and to the west in Luxeuil-les-Bains was a Turkey red works run by the Koechlin-Baumgartner branch of the family.[14] The Maison Weber firm in Mulhouse dyed good *violets façon rouge turc*, or Turkey purple mordanted with some iron.[15] Aside from the exhibition catalogue *Andrinople: Le rouge magnifique*, there have not been any detailed histories of production in these locations.

England

Printed cotton manufacturing began around London in the late 1600s, bolstered by the immigration of Huguenots leaving France. Eighteenth-century government restrictions on printed cotton products in Britain initially limited the cotton textile industry and the development of local specialist knowledge.[16] From the mid-1700s, much of the growing English cotton textile industry was concentrated around Manchester in Lancashire (see Figure 5.4), which offered space to erect new factories and an abundant water supply that powered the new machinery of the Industrial Revolution. The industry there became so expansive that Manchester earned the nickname "Cottonopolis." Although more removed from population centers of the time, the northern industry was able to compete with older London manufacturers due to the available waterpower, transportation links, and cheaper labor.

It was from Rouen that the Turkey red industry finally spread to Britain, again through the migration of knowledgeable dyers. There were some early attempts to dye it, like John Wilson of Ainsworth, (see Chapter 2), but his success in Turkey red did not become part of his established trade.[17] Credit for the first successful demonstration of Turkey red dyeing in Britain to a government

Figure 5.4 Map showing approximate locations of places in Briain associated with Turkey red dyeing. *Created with MapChart*

panel generally goes to French dyers Louis and Abraham Borelle. Louis Borelle had come to England from Rouen in late 1781 with the intent to dye Turkey red and claim a Parliamentary prize of £2,500 (roughly £200,000 in the twenty-first century). His local contact, Member of Parliament William Morton Pitt, was abroad at the time. An agent of Borelle, Mr. Crusaz, wrote to Pitt with samples in the spring of 1782 and received no reply, writing again in the autumn of 1783. At this time Pitt was about to go abroad for his health, but passed the query to Sir Thomas Egerton, who was MP for Lancashire. Again, nothing happened, this time due to the imminent dissolution of Parliament, though Egerton did take the petition to the Manchester Committee of Commerce. A request was made of the Committee secretary in early 1784 to contact Mr. Crusaz, which he did not do, but Crusaz contacted the Committee himself that June and the Borelles did their demonstration in late 1784, showing their samples in early 1785.[18] The Committee was able to replicate it "in the Hank and the Piece," and estimated the cost at £8 (about £650 today) per one hundred pounds of yarn.[19] It is worthwhile to note that piece dyeing, or whole cloth, for Turkey red was supposedly not possible at this time. If the 1810 development by Koechlin to dye cloth was based on handling the fibers, then it is possible the Borelles were also capable of it. No doubt the value of the prize sustained their patience, which endured for more than three years from arriving in England to receiving any consideration from Parliament. Finally, in 1786 they were awarded the prize. Although by his own account John Wilson was the first British dyer to make Turkey red, the Borelles are more often recognized as such because they made a formal demonstration in front of witnesses. The awarded sum was substantial but did not immediately result in a commercial enterprise, and it is unclear exactly what caused the delay. A 1786 report from the *Chambre de Commerce de Normandie* observed that unlike Rouen, Manchester did not yet have good Turkey red dyeing, but twenty French dyers had gone there and it was likely that within a year the color would be known in all of England.[20]

The first record of commercial Turkey red production by a Borelle is a *Manchester Mercury* advertisement from October 28, 1788 that said "Mr. Borel" of Blackley had a Turkey red dyeworks and was available for consultation. Local directory entries for 1794, 1797, 1800, and 1802 list Abraham Henry Borl [sic] as a Turkey red dyer in Crumpsall. The two villages, now suburbs of Greater Manchester, were adjacent with the River Irk running between them, providing the water supply. Many sources assume, though without documentary evidence, that Louis and Abraham Borelle were brothers. Louis does not appear in any documents connected to the dyeworks, but was mentioned in another 1796 Parliamentary petition for naturalization. Abraham Borelle retired in 1804 and his premises in Blackley were advertised for sale in the Manchester papers. The offer included all the tools and equipment, as well as access to Borelle who would "cheerfully render every necessary information and instruction to a purchaser." [21] A history of printed textile manufacturing in Neuchâtel says that the Borelles who demonstrated Turkey red dyeing in Manchester were from a branch of the Borel-de Bitche family that migrated to Normandy sometime in the eighteenth century.[22] The family had operated a textile firm in Couvet, between Lake Neuchâtel and the French border, from 1750 to 1772. A history of the family lists multiple members named Abraham Henri and Louis, though none that are brothers.[23]

The Borelles were not the only prospective Turkey red dyers in Lancashire. Charles Taylor, a leading Manchester dyer and calico printer, brought a Greek Turkey red dyer to England sometime before 1788, when Taylor referenced the endeavor in a letter to James Watt, the Scottish engineer. Watt's son, James junior, was apprenticed to Taylor's firm Maxwell, Taylor & Company and also mentioned Turkey red in a 1790 letter to his father. One source gives 1785 as the year Taylor started Turkey red dyeing in Manchester, and that he was probably using the Borelle process with improvements based on novel chemical developments.[24] According to this, Taylor would have been the first commercial Turkey red dyer in England and went into business before the Borelles received the Parliamentary prize, though it is not explained how he would have been following their process. Another source, however, remarks on the absence of advertisements for Turkey red yarn from Maxwell, Taylor & Company from this time. The earliest notices in the *Manchester Mercury* mentioning Turkey red appear in 1787, though this does not eliminate the possibility Taylor's firm was dyeing before then. In March 1791, a Greek Turkey red dyer named Theokary, formerly employed by Taylor, advertised in the *Manchester Mercury* that he was leaving the country and would be at the Gibraltar Tavern "if any Person has any Demands upon him" for the next three weeks. It may be that Taylor brought Theokary to Manchester, but they were not successful commercially, and that this happened around the same time the Borelles were petitioning for their prize.

An advertisement in the *Manchester Mercury* on June 24, 1788, four months before the one by Borelle, says "Delaunay and Payant beg leave to inform their Friends and the Public that they have now completed their Dye-House, etc., at Crumpsall, and are ready to take in Hank Cotton for dyeing the permanent Turkey Red, to return it in three or four Weeks." Angel Raphael Louis Delaunay was a French dyer from Rouen, around thirty-five years of age, who had immigrated with his wife and their two young children. He was partnered with Charles Payant, another French immigrant, who had been in Manchester teaching French and Italian since 1777. The partnership did not last long, and Delaunay advertised in 1790 that he was no longer working with Payant but still dyed Turkey red and "other fine Colours." In 1794, Delaunay moved his dye house to Blackley. He filed for bankruptcy in 1808, possibly due to the economic effects of the Napoleonic Wars, and his assets were liquidated with the debts settled before his death in 1811. There was enough of an estate to leave his eldest son Louis a third of his property and the wish that he continued the dyeing trade. Louis operated a dyeworks in the area from 1811 until his death in 1865. He was regarded as a good employer and expanded the firm to include multiple dyehouses, a printworks, administrative buildings, and gas infrastructure for lighting the premises. The nearby road he laid survives today as Delaunays Road and runs past the North Manchester General Hospital.

Regardless of who established the Turkey red industry in Lancashire, there were multiple parties working toward commercial production from the 1780s onward. James Thomson, the textile manufacturer who received an English patent for discharge printing, dyed Turkey red at his Primrose works near Clitheroe. Thomson, a talented chemist, was another friend of James Watt and has been called the most celebrated textile printer of the first half of the nineteenth century.[25] It is not clear when Thomson started dyeing Turkey red, but his knowledge, experience, and connections no doubt facilitated his work. Another friend of Thomson, the Alsatian dyer Frédéric Steiner who developed the hot oiling technique, studied with Daniel Koechlin in Mulhouse and visited Primrose, where he learned about Turkey red dyeing in the region.[26] Steiner worked for a period at the Broad Oak printworks in Accrington, which was owned by the Hargreaves family, and in 1836 purchased part of an establishment near Church, owned by the famous Peel family of textile printers, to open his own Turkey red firm.[27] Foxhill Bank printworks was located in the valley of Tinker Brook, west of Union Road in Oswaldthistle. It was founded around 1780 and changed hands a few times in the mid-nineteenth century. A combination of machine printing and hand blocks were used at Foxhill Bank, which also dyed Turkey red. The site was purchased by the Steiner firm in 1892 and it closed in 1931.[28] It is not clear when Turkey red production began, but a manuscript from the 1830s to 1840s in the collection of the Victoria and Albert Museum contains notes on Turkey red dyeing along with a sample.[29]

There may be other, smaller dyers of Turkey red that were operating in Lancashire, but the firms discussed here were the major manufacturers in the English industry. Figure 5.5 shows the halls of the Crystal Palace in London decorated with Turkey red cloth for the 1851 Great Exhibition, showcasing its significance to British manufacturing. An entry in *The Record of the International Exhibition, 1862* said that yarn was still purchased in Manchester and sent to the Netherlands for dyeing Turkey red before being returned to Manchester for sale with English-dyed yarn, implying demand still exceeded local production capacity.[30]

Figure 5.5 A lithograph from Dickinson's Comprehensive Pictures of the Great Exhibition of 1851 by Joseph Nash showing the Crystal Palace draped in Turkey red. *Cooper Hewitt, Smithsonian Design Museum. Gift of Ashton Hawkins. 1998-65-1*

Scotland

In the context of Turkey red, a distinction between the Scottish and English industries is often made because of the sheer scale of production in the Vale of Leven, a region about 20 miles (32 km) northwest of Glasgow bordering the River Leven, which flows from Balloch at the southern end of Loch Lomond into the Firth of Clyde. An account of the International Exhibition of 1862 held in London observed that all samples of Turkey red shown came from Scottish manufacturers rather than English ones, and that they were "by far the best in the Exhibition."[31]

Political events in the eighteenth century led to the establishment of a robust and technologically progressive Scottish textile industry, starting with the unification of Scotland and England in 1707. This opened the American market to Scottish manufacturers. In 1701, Scotland owned 215 ships, a figure that increased more than fivefold a decade later. Agricultural reforms and the Industrial Revolution brought prosperity to the nation. The Highland Clearances, large-scale evictions of tenant farmers that began in the mid-1700s, pushed a large pool of laborers south to the industrializing Central Belt of Scotland, providing many able hands. Wool and linen had been produced in Scotland for some time, and its abundant soft water was ideal for textile dyeing and printing. In 1738, linen printing was introduced by Archibald Ingram, called the father of Scottish textile printing. Ingram was also one of Glasgow's Tobacco Lords, the eighteenth-century merchants whose fortunes were made trading in North American tobacco and sugar. These economies were intertwined with the labor of and trade in enslaved people on the triangular route between Europe, Africa, and the Americas, where textile goods were also exchanged. In the late 1770s, the American Revolution caused the collapse of the tobacco trade, a loss felt economically in Glasgow. Local merchants adapted to market changes by investing in textile manufacturing, which was expanding with the progression of the Industrial Revolution. Rising interest in dyeing led to the creation of groups like the Red Society, which was formed with the intent of "encouraging the dyeing of Mather Red" on cotton and linen.

One of the first large dyeing establishments in Glasgow was founded in 1777 by George Macintosh (also written as Mackintosh or MacIntosh). He had been born in the Highlands and moved to Glasgow as a young man to seek employment in manufacturing. Macintosh worked in a shoe factory before securing an investment from local merchants to open a cudbear dyeworks, where he employed many Highlanders driven south by the clearances. Cudbear, a reddish-purple dye made from lichens, was particularly noxious because it required ammonia, which was recovered from human urine. The works consumed about 2,000 gallons daily, collected from iron casks "dispersed among the manufacturing and tradesmen's houses in Glasgow and suburbs." The annual cost of this collected urine was estimated to be £800, roughly £67,000 in the twenty-first century. Cudbear was good for dyeing silk and wool, but not very effective on cotton, which had the real potential for industrial growth.

Most of the textile industry was located in the western part of the Central Belt, around Glasgow where there was already some infrastructure and in the Vale of Leven northwest of the city. The Vale had an abundant water supply and ample open space along the riverbanks where fabric could be spread for bleaching and drying. Experienced Dutch immigrant workers had begun textile bleaching on the banks of the Leven around 1715. They designed channels dug from the river to bring water into adjacent fields, and planted beech hedgerows for windbreaks to protect the drying cloth. In 1727, William Stirling, a surgeon in Glasgow, founded the Dalquhurn Bleaching Company on the Leven. The first printworks was founded around 1768, and two more were in operation by the 1790s. They were early adopters of water-powered machinery, which significantly increased production capacity.

In early 1785, a dyer from Rouen named Pierre-Jacques Papillon wrote a letter to a member of the committee for the same Parliamentary prize that the Borelles received. For unknown reasons, he used the pseudonym Cigale, which means "cicada" in French, possibly a play on his surname which means "butterfly." Satisfied with the work of the Borelles, the committee did not take him up on the offer. Although he was thwarted in claiming the £2,500 prize, another opportunity arose. While in London that year, Papillon met George Macintosh, who brought him to Glasgow. Along with Macintosh's partner David Dale, they established the Dalmarnock Turkey red dyeworks on the banks of the River Clyde in late 1785, at least a year before the first advertisement for Turkey red in the *Manchester Mercury*. The works, long since demolished, were in the area between present-day French Street and the Clyde. An advertisement in *The Glasgow Mercury* on December 15, 1785 said "Dale and Macintosh have now got their Dyehouse finished, and are just begun to dye cotton yarn Turkey Red … The excellency of this colour is already known here, as it has been tried and found to stand the process of bleaching, when woven along with green linen, or cotton, yarn, without impairing, but rather increasing its beauty and lustre." The dyers requested consignments of at least sixty pounds of yarn to be dyed for three shillings per pound (roughly £12 today). A quantity of dyed yarn was also kept for immediate sale at a slightly higher price, but initially those wanting Turkey red yarn had to supply their own for dyeing. Conveniently, Dale had co-founded a cotton mill called New Lanark in 1783 in partnership with Richard Arkwright, the inventor of significant spinning and carding machines, and could provide an ample yarn supply.

A 1787 letter from George Macintosh to his son Charles, who invented the waterproof raincoats called Macintoshes, described the departure of Papillon from the firm as they "could not manage his unhappy temper" and that his process was since refined so dyeing could be completed in twenty days instead of twenty-five. Papillon, meanwhile, set out to obtain funding from the Board of Manufactures to start his own works, which likely would have been refused were it not for the support of the Lord Advocate, the chief legal adviser to the British government on matters of Scottish law. Papillon received a grant of £100 (about £7,700 in the twenty-first century) and an award of £50 annually for the next two years. As collateral, he was required to write out his process, which was to be sealed for the duration of their agreement. It was examined by the chemist Dr. Joseph Black and entrusted to him for the embargo period. In June of 1787, Papillon advertised his dye house, which was located fairly close to Dalmarnock, in *The Glasgow Mercury*, where he claimed to dye Turkey red superior to any other in the country. Despite the financial support and some initial commercial success, for unknown reasons Papillon could not sustain his operations and eventually died in poverty and obscurity. The embargo on his method ended in 1802, and the method was published in *The Philosophical Magazine* in 1804.

1785 was a significant year for Scottish Turkey red and David Dale. Besides Dalmarnock, he established a cotton mill in Blantyre, southeast of Glasgow, with another business partner, James Monteith. Blantyre soon became the second place in Scotland where Turkey red was dyed, and was one of few places where cotton spinning and Turkey red dyeing were done in the same location. It became common from the late eighteenth century, as factories expanded, to provide employee housing on-site or nearby. Blantyre's most famous son, the missionary explorer David Livingstone, was born in the workers' tenements in 1813 and as a child worked as a cotton piecer in the mills tying broken ends on spinning machines.

Dale sold his shares to Monteith in 1792. In 1802, his younger brother Henry Monteith took over operations and also founded the Barrowfield weaving factory of Henry Monteith, Bogle & Company in 1802, which produced yarn-dyed bandanas (and later the famous discharge-printed ones). Henry became a major manufacturer of these bandanas and of pullicates, and in 1805 purchased Dalmarnock from Dale and Macintosh. Dyeing at Dalmarnock, which was subsumed under the Barrowfield name, continued under Henry's son Robert until his retirement in 1873, after which it was demolished. The Monteith firm was liquidated in 1904 and in 1925 the Blantyre site was acquired by a trust, now open as the David Livingstone Birthplace Museum.

Other smaller Turkey red firms existed in the Glasgow area. In 1796, Peter Ferguson founded a bleachfield at Shawfield Bank south of Glasgow. His successors, the Gowdies, introduced Turkey red dyeing there, though it is not clear for how long it was made. A Mr. Mathieson built a Turkey red works near Rutherglen in 1833; its longevity is also indeterminate. There was also T.P. Miller & Company in Cambuslang, and J. & W. Campbell in Pollockshaws. Walter Crum operated his family textile firm in Thornliebank which also made Turkey red. He had studied under James Thomson of Primrose and later becoming a Fellow of the Royal Society. By the early nineteenth century, the increasing population of the city and corresponding pollution of the water supply pushed the larger operations for dyeing and printing westward out of the city toward the Vale of Leven.

Turkey red was first dyed there around 1827 at the Croftengea works, which had been founded some years earlier as a bleachfield under John Turnbull and his partner. Turnbull received the site in 1802 as part of a deal for services rendered to his former employer, William Stirling and Sons. That firm had been established by the nephew of Dalquhurn Bleaching Company founder William Stirling, who shared his name. The younger William began his career selling imported Indian cotton fabrics printed in London. By 1750, he had opened his own printworks on the River Kelvin at Dawsholm in the northwest of Glasgow. Due to the diminishing water quality, he relocated to the Leven in 1770 and leased land from John Campbell of Stonefield. Together with his sons, Stirling built Cordale printworks and formed William Stirling & Sons. Stirling died in 1777, and the company continued to expand under his heritors. In 1791 they acquired Dalquhurn, which did the bleaching and dyeing to supply Cordale with fabric for printing. In 1859, Dalquhurn dyed 18 million yards (16.5 million meters) of Turkey red fabric and 800,000 pounds (360,000 kg) of Turkey red yarn. Around the same time, Cordale was called the most extensive Turkey red printing operation in Scotland. It occupied about 70 acres (28 hectares), ten of which were buildings. The scale of the industry was such that in the first half of 1882, William Stirling and Sons was valued at around £208,000, the equivalent of roughly £14 million in twenty-first-century currency.

The first printworks on the Leven, called Levenfield, was founded in 1768. It was owned by John Todd and Company, who also founded a bleachfield called Milton across the river in 1772. A ferry connected the two sites, carrying goods and workers. John Todd and Company operated Levenfield until 1850, when it was sold to John Orr Ewing, a partner in a Glasgow yarn business. Orr Ewing had leased Croftengea from William Stirling & Sons since 1835 with partner Robert Alexander, operating under the name John Orr Ewing & Company. Business expanded rapidly with the assistance of a colorist named John Wylie, enabling the purchase of additional land to expand the works in 1840. The site was renamed the Alexandria Works and made printed cottons, eventually amalgamating with Levenfield. In 1859, following an industrial trend, the firm hired the first chemist employed in the Vale, John Hyde Christie.

Archibald Orr Ewing, the younger brother of John, started his own business in 1845. He had managed Croftengea for his brother, becoming familiar with Turkey red dyeing and printing. Together with partner William Miller and financial backing from John, he formed the Archibald Orr Ewing firm. They began leasing the Levenbank works, which had been founded as a printworks in 1774. Archibald acquired the Milton site from John Todd and Company in 1850 and enlarged it for dyeing Turkey red yarn. The brothers quarreled in the early 1850s, causing John to withdraw his investment in Archibald's firm. Undeterred, he purchased Levenbank outright in 1853. In 1866, Archibald Orr Ewing acquired the Dillichip works, which had been founded sometime before 1839 based on its mention in the *New Statistical Account of Scotland*.

The textile industry expanded rapidly from the mid-nineteenth century with the growth of local infrastructure, which benefited from the proximity of coal mines in Lanarkshire. In the late nineteenth century, the North British and Caledonian Railway companies had a branch of their system running into the Dalmonach printworks, which operated twenty-eight print machines and produced more than 25 million yards (23 million meters) of cloth annually as well as dyeing more than one million pounds (450,000 kg) of yarn. William Stirling and Sons, John Orr Ewing and Company, and Archibald Orr Ewing and Company became the three main Turkey red dyers and printers in the Vale of Leven. Demand was great enough that all three were successful throughout the late nineteenth century, the "heyday" of Turkey red. The scale of the industry in the Vale, referred to locally as "The Craft," was such that it was present in nearly every aspect of workers' lives. Collectively, the firms were the single largest employer in the area and owned much of the local property. Factories operated around the clock except for Christmas and New Year's Day, and nearly every family had at least one member employed in textile dyeing or printing. One man recounted his childhood playing on the banks of the Leven in the early twentieth century, saying "[W]e jumped back from the odd jets of steam which spurted from its banks and were drawn to the gaping sluice gates which took giant gulps of fresh water and spat it out—blood red."[32]

In 1897, newly-imposed tariffs by the Indian government reduced market access. The corresponding decrease in revenue pushed Archibald Orr Ewing, John Orr Ewing, and William Stirling and Sons to amalgamate as the United Turkey Red Company. This operated as an umbrella firm, with the constituent manufacturers continuing to make their own products. United Turkey Red included the works Dalquhurn, Alexandria, Levenbank, Cordale, Milton, and Dillichip. In 1900, Alexander Reid of Milngavie also joined. The first chairman was John Hyde Christie, the chemist hired decades earlier by John Orr Ewing. Christie served as chairman until 1922, when he retired at age eighty-six. He was briefly succeeded by William Ewing Gilmour, who died after two years and was replaced by Christie's son Harry. Harry's proved to be unpopular with the workers at a time when other developments in the textile industry were causing the organization to suffer. Synthetic red dyes had finally improved to the point of becoming competitive with Turkey red, further threatening the market. Turkey red dyeing in the Vale of Leven ended in 1936, but United Turkey Red continued as a textile firm making other products until 1961 when they were taken over by the Calico Printers Association and closed down. United Turkey Red produced many sample pattern books like the one shown in Figure 4.17. Company documents and shipping manifests show Scottish Turkey red was traded to India, South America, the Middle East, Indonesia, China, the Philippines, Africa, the South Pacific, and Japan. The success of this was due in large part to the advantages of existing networks that formed during the tobacco trade. A more detailed study of the Scottish Turkey red industry presented through the lens of the sample pattern books held by the National Museums Scotland can be found in Nenadic and Tuckett's *Colouring the Nation: The Turkey Red Printed Cotton Industry in Scotland c. 1840-1940*.[33]

Switzerland

There is no comprehensive history of the Swiss Turkey red industry, but it is discussed some in more general histories on Swiss textile manufacturing which are mostly written in French and German. Textile production in Switzerland benefited immensely when, in consecutive years, the French government outlawed Protestantism and prohibited printed cottons to protect domestic wool and silk industries. The subsequent exodus of skilled Huguenot textile workers benefited the Swiss and German provinces where they re-established their trade.[34] Restrictions on printed textiles were not as prevalent in Switzerland, fostering more innovation. The relative economic freedom, along with the arrival of skilled immigrants, were the perfect conditions to support a growing cotton textile industry. Switzerland was a major manufacturer of printed cottons in the eighteenth and nineteenth centuries. Early establishments appeared in the first decades of the eighteenth century and were concentrated in the cantons of Neuchâtel and Glarus (see Figure 5.6). The focus was on dyeing, printing, and finishing, and less on processing raw cotton. Neuchâtel was supplied with bleached and unbleached cotton cloth imported via overland trade routes from India by Dutch, French, and English traders. Early training for textile printers in Switzerland took place in Geneva, with much of the knowledge coming from Huguenot immigrants and the more established Dutch industry.[35] There was a continuous exchange of information and labor between the textile industries of Switzerland, Mulhouse, and France during this period, and travel further abroad to Britain. By the mid-eighteenth century, when Turkey red dyeing was established in Western Europe, Swiss cotton textile manufacturing had the infrastructure and skilled workers that such a specialized operation required. Turkey red printing was prevalent in the cantons of Zurich, St. Gallen, Thurgau, and Glarus, the last of which exported a large quantity of prints around the world, including the Balkans, North Africa, India, and Malaysia.

Multiple firms have been designated the "first" Turkey red dyer in Switzerland, probably based on what records and documents the author could access. The earliest mention of a Turkey red works is from 1784 and it was located in Drahtschmidli, part of Zurich. Another source says Turkey red dyeing was introduced to Switzerland by Johann Heinrich Zeller of Zurich, who worked for a silk dyer in Nîmes, France (near Marseille) as a young man where he learned the technique before returning to Switzerland and founding a works near Zurich. Based on the two sources, it may be

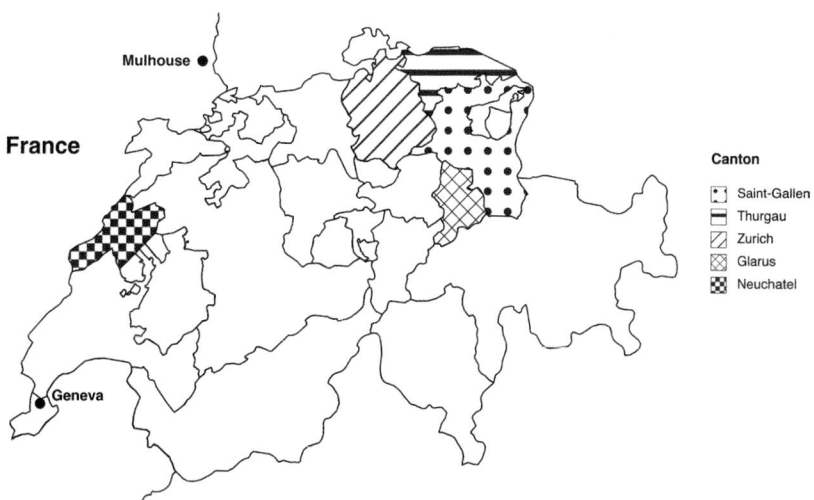

Figure 5.6 Map showing approximate locations of places in Switzerland associated with Turkey red dyeing. *Created with MapChart*

Zeller who started Turkey red dyeing there in 1784. In nearby Winterthur, the Sulzer firm, founded by Hans Heinrich Sulzer, had a dye and print works that made Turkey red until 1830, when they closed the business due to decreasing local water quality and some business conflicts.

Johann Heinrich Sulzer moved to Aadorf, just over the border in the canton of Thurgau, in 1833 and took over a spinning mill. He was married to Maria Luise Steiner, a possible relation of the Ribeauvillé Steiners. The industry in Thurgau started in the 1790s when Luthens and Rosier dyed yarn in Frauenfeld on the Mühle Canal, though they had little success due to the inadequate quality of the product and the firm closed within a few years. The Sulzers began dyeing Turkey red and his son, Jakob Heinrich, studied with the chemist Justus Liebig and implemented many technological improvements when he took over the family firm. The Greuter firm in Islikon (now part of Gachnang) started dyeing Turkey red in 1830. The firm later expanded into Alsace, which at the time was politically independent from France and Germany, to avoid customs regulations. Elsewhere in Thurgau, Turkey red was dyed in Hauptwil, Islikon-Frauenfeld, Güttingen, Bürglen, Arbon, Amriswil, Bischofszell, Luxburg, and Staubishub.

Just to the west in the canton of St. Gallen, a Turkey red works was founded in Uznach in the 1820s by Benedikt Schubiger and his partners. It was sold to Gottfried Hofmann in 1875. The Sulzer family acquired a Goßau firm founded by J.J. Kelly that dyed and printed Turkey red. Figure 5.7 shows an 1850 watercolor of the factory by the artist Elizabeth Kelly, daughter of the founder. The red-draped drying towers (like Figure 3.5) were a common feature of Swiss Turkey red manufacturing, which took advantage of a more suitable climate for outdoor drying and saved the cost of heating ovens.[36] In the canton of Glarus, the first Turkey red was dyed around 1817 by Egidius Trumpy, whose firm had been founded some twenty years earlier. In Schwanden the Tschudi firm, founded by Johann Caspar Tschudi in 1829, also dyed Turkey red, as did the Jenny and Blumer firm.[37] Although there are no examinations of the Swiss Turkey red industry like

Figure 5.7 An 1850 watercolor of the Swiss Turkey red works of J.J. Kelly painted by his daughter Elisabeth, showing red cotton hanging from the drying tower. *Turkey Red Printing Manufacture J. J. Kelly, Gossau, Switzerland, about 1850 by Anna Elisabeth Kelly*

Colouring the Nation does for the Scottish industry, there may be sufficient records of firms and production in archives and museums that, along with objects, such a story remains to be told.

The Netherlands

Into the early eighteenth century, textile finishing, particularly for dyeing and bleaching, was practiced at a higher level in the Netherlands than in Britain.[38] Like Switzerland, the Netherlands did not have as many legal restrictions on printed textiles compared to France, Germany, and England, allowing more development.[39] In 1678, madder dyeing was introduced in Amersfoort by two Amsterdam merchants who recruited an Armenian dyer named Lowijs de Celibi through intermediaries.[40] As discussed in Chapter 2, madder became a significant commercial crop in the Netherlands and the growing local textile industry helped develop more knowledge about its use.

Some accounts mention an eighteenth-century Turkey red workshop in Leiden (see Figure 5.8), established around the same time as the first in Rouen in 1747.[41] It is mentioned in a 1767 French text, but there does not seem to be any further evidence of this establishment.[42] In 1778, a competition was organized by the Oeconomische Tak van de Hollandsche Maatschappij der Wetenschappen, the economic branch of the society for the promotion of sciences, to establish production of "Arabian Red." It was renewed annually in the absence of entries until 1800, when a manufacturer named Van Baer from Eindhoven submitted some samples and was awarded the prize. By 1801, there were two Turkey red dyers in Eindhoven, but business did not survive the Napoleonic Wars. Deussen and Röder, two dyers from Gladbach in Germany, founded a Turkey red firm to the northeast of Eindhoven in Aarle-Rixtel in 1829. P.F. van Vlissingen founded a textile firm nearby Helmond and employed a Mulhousian chemist named Braun who later went on to found his own Turkey red factory in Vlaardingen. Later, Helmond was also the site of firms Kaulen and Carp, and Swinkels.

In the 1830s, textile production in the Netherlands suffered with the secession of Belgium as an independent nation, where much of the industry was located. Manufacturing was rebuilt with the help of foreign entrepreneurs and technicians. The Nederlandsche Handel-Maatschappij (Netherlands Trading Society), established in 1824 to promote and develop trade, shipping, and agriculture, worked to revive the textile industry by bringing factories to the Netherlands. More Turkey red firms making yarn and cloth were established during this time. In Beverwijk, Prince & Co. was dyeing Turkey red by 1833. The next year, Belgian textile manufacturer J.B.T. Prévinaire opened a factory in nearby Haarlem. Prévinaire was an expert Turkey red dyer, and his firm became

Figure 5.8 Map showing approximate locations of places in the Netherlands associated with Turkey red dyeing. *Created with MapChart*

a significant part of the Dutch industry. He had learned the method as a young man in Rouen, then established a Turkey red factory in Molenbeek-Saint-Jean near Brussels. Although he brought two experienced dyers with him, the Haarlem operation struggled initially due to a lack of specialized labor, so he recruited more foreign workers. Quality still suffered for a period due to the local water supply and other factors, so Prévinaire illicitly imported his own Turkey red from Belgium for a period to compensate while adapting at the Haarlem site. He did not receive any fines for this, but his monopoly was revoked and support was given to other dyers. In 1852 the Jannsens brothers, sons of a pharmacist, opened a Turkey red operation in Herten. Pharmaceutical experience somewhat compensated for the lack of technical chemical knowledge, and was an advantage in dyeing.

During the period immediately following the Belgian secession, dyers in the Netherlands relied on French and German publications for knowledge to revive the textile industry. De Heyder, a cotton printing company that relocated to Leiden from Belgium in 1836, also produced Turkey red yarn and cloth in the latter half of the 1830s and much of their skilled staff was Belgian.[43] The De Heyder firm was acquired by the Driessen family later in the nineteenth century, becoming the Leidse Katoenmaatschappij, where they continued to dye Turkey red. Heinrich Driessen was a textile merchant rather than a dyer, and tasked his sons with reviving the firm. One of the resources used to accomplish this was *Traité théorique et pratique de l'impression des tissus* by Jean-François Persoz (see figures 4.11 and Figure 4.12). One brother, Pierre, was sent to Rouen to learn the trade as a colorist. Another, Félix, trained in Mulhouse, becoming a talented colorist and developing professional connections through the Société Industrielle de Mulhouse. He later became the head of the firm, and traveled to Indonesia to study batik and determined the role of aluminium-accumulating plants like *Symplocos* in local Turkey red practice. Driessen used his knowledge to produce Turkey red batiks at the Leiden factory, though it is not clear how these were made.[44] The firm became successful, and in an effort to reduce dependence on foreign experts he built and maintained detailed internal documentation on production, some of which can be found in the collection of the TextielMuseum in Tilburg. A former associate of Driessen named Van Wensen left the firm and later dyed Turkey red yarn in Zoeterwoude, south of Leiden.[45] It is not clear from the literature to what extent Turkey red was dyed in the lands that became Belgium.

North America

European colonization of North America began in the early seventeenth century, but the demands of survival precluded the development of a specialized industry like textile dyeing and printing. Britain maintained the largest colonies, but France, Spain, and the Netherlands also had an active presence in the Americas. Although settlements eventually stabilized and the population grew enough to support manufacturing, colonies remained dependent on Atlantic trade since it was in the interest of British manufacturers to keep their captive market for British goods, one of the complaints that fomented the American Revolution. As such, there was little industrial textile manufacturing until after the war. A 1771 export list from Glasgow says "twelve pieces of 'Bandannoes' went to Virginia and one piece to North Carolina," presumably for enslaved people in the colonies. This account pre-dates Scottish Turkey red dyeing, but indicates pre-existing trade links through which it would be exchanged within the next two decades.[46]

Efforts to develop a textile industry in the new United States of America were hindered by British efforts to protect market dominance, which included a ban on exporting textile machinery and one on immigration for men with technical knowledge of textile manufacturing. A few had managed to relocate prior to the Revolutionary War, and traveled to Philadelphia, Pennsylvania and Providence and East Greenwich in Rhode Island.[47] Toward the end of the eighteenth century, sheep were kept in the northern parts of the United States and some wool dyeing was practiced. Asa Ellis, the author of one of the earliest American texts on dyeing, observed that "few people, in America, estimate the value of manufactured woolens, of their own country. We too generally resort, for our cloths, to the manufactories of Europe."[48]

Until the mid-nineteenth century, most professionals in the industry were either European immigrants or had received training in Europe. They tended to prefer ingredients familiar to the European industry, which were often also of superior quality since the American market had not yet recovered following independence. Prominent Americans like Thomas Jefferson and Dolley Madison promoted domestic cultivation of dye plants. In 1785, the Society for the Promotion of Agriculture in South Carolina offered a premium for growing madder, which became a personal interest of Jefferson. An 1811 note in his garden book shows he planted imported *Galium mollugo*, a relative of *Rubia*, from seeds obtained in France. Jefferson also incorrectly claimed fresh madder root had twice the strength of dried, but the desire for knowledge was present.[49] Mills for spinning and weaving were built throughout the northern states in what is now called New England. An 1806 American dyeing manual encouraged the promotion of American commercial interests to support independence from Britain and other foreign countries, "to whom in arts and manufactures we have too long bowed the knee."[50]

Even after the development of an American textile industry, Turkey red was mostly imported from Europe until the 1860s.[51] The complexity of the technique, especially with the old process, hindered manufacturing in the United States where the industry remained less technically advanced than in Europe.[52] There is evidence of small-scale production, or at least knowledge of the technique, from the early nineteenth century. An 1811 article by Lawrence A. Washington describes his wife's success in dyeing Turkey red based on a process in *The Domestic Encyclopedia*. This was first published in London in 1802, and was followed by an American edition printed in Philadelphia in 1803, likely what Mrs. Washington read. Washington observed the United States had a "deficiency . . . respecting the art of dyeing" and authored his article with the intent of contributing to domestic knowledge on the topic. He offered to provide a simplified recipe in a later article since the book was not commonly available.[53] In 1815, *John Rauch's receipts on dyeing, in a series of letters to a friend*, was published in New York. Rauch was a Swiss immigrant and dyer who worked in Massachusetts, Connecticut, New York, and New Jersey in the early nineteenth century, traveling around providing instruction in dyeing for a fee. This printed edition is possibly the first American publication that is not a reprint of a European text to contain a Turkey red method. Rauch coded the quantities of ingredients in the published version and required additional payment for the key. Interestingly, the manuscript version with handwritten processes by Rauch does not contain a Turkey red method but does contain a key to his coding system.[54] Another possible first, also published in 1815, was a full dyeing treatise by Thomas Cooper that also contains a method for Turkey red.[55] Cooper came from Manchester, and was a pioneer in the field of bleaching and dyeing who had emigrated to the United States. He advocated for the advancement of chemical research, knowledge, and theory to support the art of dyeing, and his book was considered to be the most important early publication on the topic in the country.[56]

The American textile industry expanded during the early nineteenth century as American textile manufacturers sent representatives to Britain in an effort to recruit skilled workers, chemists, engravers, printers, and colorists.[57] In turn, European dyeing specialists were attracted to the United States for the better wages paid.[58] Turkey red was dyed and printed in the second half of the century, and the works discussed here were documented to have made it during this time. It is a period with limited available records, so often only names and locations are known. Production was concentrated in New England, where the first spinning and weaving mills were built in the early nineteenth century. In Massachusetts, the Hamilton Print Works was founded in 1825 in Lowell and in the late nineteenth century made Turkey red bandanas.[59] Evidence suggests they also dyed Turkey red with madder, but quickly transitioned to synthetic alizarin when it became available.[60] The Cochrane Turkey Red works, which operated in Malden (now part of Greater Boston), was established in 1857. John Cochrane had emigrated from Scotland about a decade earlier and had some knowledge of the textile trade from his father. The firm operated until the 1920s and produced many textile items in addition to Turkey red, but they are particularly known for their commemorative handkerchiefs made for major events, political campaigns, and other such occasions.[61] In 1876, the Walpole Dye and Chemical Works (conveniently situated about halfway between Boston and Providence) began selling Turkey red oil. The former superintendent of the works, a Mr. Lane, claimed they were the first to sell it, but established dyers who had emigrated from Scotland said they were making their own supply before it was commercially available.[62]

The Cocheco Print Works, founded in Dover, New Hampshire in 1836, used French, Dutch, and Turkish madder, garancine, madder extracts, and synthetic alizarin. Cocheco was one of the earliest large-scale textile printing firms in the United States and produced engraved roller prints of patterns similar to those made in Britain.[63] The Clyde Bleachery and Print Works was founded in 1828 in West Warwick, Rhode Island by Simon H. Greene and his partner Edward Pike. They became famous for their "Washington" line of prints which invoked Martha Washington, the First Lady of the United States, made in the later nineteenth century.[64] The name of the works implies a Scottish connection with the River Clyde that flows through Glasgow, but whether there is a link to Scottish Turkey red is unclear. Robert Reoch, the superintendent of the Clyde Works in the mid-nineteenth century, said it was not until after 1869 and the introduction of synthetic alizarin that Turkey red was printed in the United States. Reoch was hired from England in 1867 and remained at the site until the early 1900s. Bandanas were printed following the Monteith discharge method of clamping the fabric between lead plates.[65] Henry Ashworth founded the Turkey Red Dyeing Company in 1876 in Providence, Rhode Island.[66]

In a 1916 book on textile dyeing in Germany and America, the author comments that despite what was said about American manufacturers being progressive, they had "not much to teach us as regards the dyeing industry." The consumption of textiles in the United States was so great since "different seasons of the year demand entirely different clothes" that manufacturing standards were lower. The inherent difference in the market, along with the tendency for American manufacturers to be responsible for the entire process from spinning to finishing rather than contracting out certain aspects to a bleacher or dyer as in Britain, meant the aim in the United States was "turning out the stuff cheaply and quickly." As a result, there was little drive to build an export market.[67] Turkey red was certainly dyed and printed in the United States, though never on the same scale as it was in Europe. Surviving textiles indicate it was primarily discharge-printed bandanas, typically red and white with added black. The elaborate, multi-colored prints such as those made in Glasgow and Mulhouse do not appear to be a major American product, though the gaps in surviving records make this difficult to determine conclusively.

Other Locations

Turkey red was dyed in other places, although less documentary evidence survives, and the scale of the industry was not as extensive as the locations already discussed. There appear to be few operations in present-day Germany, known as Prussia until the late nineteenth century, though this may not present an accurate history. There is no comprehensive study of Turkey red dyeing in Germany (or Prussia), although there is likely information dispersed in other publications and archive material. In the early nineteenth century, Turkey red was dyed in Mariakirchen, located on the Kollbach river in southeastern Germany near the Austrian border.[68] A Turkey red works was located in what is present-day Wuppertal.[69] Some sources say cotton from the Caribbean was sent to Leghorn (Livorno, Italy) for dyeing, although this may only have been a forwarding port to the Levant rather than a place where Turkey red was dyed.[70] Some dyeing also persisted in Austria-Hungary. F. Hiller and Blaschka & Company exhibited Turkey red yarns at the 1862 International Exhibition,[71] and it was also made by Rosenthal at Hohenelbe in present-day Czechia.[72]

Turkey red was introduced to Russia by dyers from Bukhara, in present-day Uzbekistan. It was practiced in Tatar villages Ura and Urabashak, where the textiles were called *burlats*. It is not clear from the names where these villages were located or whether they still exist. From the late eighteenth century, it was also dyed in Kazan and Astrakhan. These operations may have supplied the yarn that was embroidered on the decorative border shown in Figure 2.7. Later firms were operated by the Prochoroff brothers, the Tretjakoffs, a Manuiloff, a Konshin, and an Emil Zundel.[73] In the Moscow area, Turkey red was dyed by Asaph Baranoff & Company, the Baranoff Manufacturing Company, and Zahar Morosoff Jr. Many skilled Alsatian workers went there after Alsace came under German rule following the Franco-Prussian War. Most of the goods were sold in Russia, Asia, Siberia, and Iran.[74]

Working Conditions and Labor

From the early days of industrialization in the late eighteenth century until labor rights movements in the early twentieth century, a working week in the textile industry was significantly longer than the standard forty hours of today. Shifts were often twelve hours long and Saturday was a typical workday, though maybe with reduced hours. Prior to the introduction of child labor laws, which strengthened in the twentieth century, it was common for young children to also spend long weeks in the factories doing less skilled, but no less dangerous work. Working conditions were also frequently unregulated regarding health and safety, or else regulations were unenforced. Some employers provided amenities for their staff, including nearby housing or a schoolhouse for younger children that older ones could attend part-time in the evening (after a twelve-hour day). A report from the Factories Commission in 1834 summarized survey responses from various manufacturers, including Henry Monteith for the Blantyre works. The response said that managers and overseers frequently opened windows to increase ventilation, but that they were closed by workers when possible. The report does not explain why, however considering the Scottish climate it would have been cold, wet and windy to have open windows much of the year. Dangerous parts of machinery were fenced off, and although children rarely worked in the dyehouse it did have two ten-year-olds, an eleven-year-old, and a fourteen-year-old working in 1833. The youngest mill workers were nine, and the youngest weavers ten—children under twelve were better in the spinning mill because their size enabled them to clean the machinery. Boys also assisted block printers by working as tearers, in close contact with the chemicals. All workers, regardless of age, worked twelve hours for five days and nine hours on Saturday, with breaks of forty-five minutes for breakfast and sixty for dinner, for a working week of approximately 70 hours.[75] An 1835 account of Blantyre said the shifts were from 6:00 a.m. to 7:45 p.m., and that the site employed 362 men and 553 women. There was a chapel, for which the company paid half the minister's fees, and a schoolhouse, where an average of 136 children attended daytime courses and fifty-six attended evening courses. The company kept the surrounding village where the workers lived clean and neat, and provided a public wash house and water pump.[76] Educational opportunities were also sometimes provided for adult employees since skilled labor was increasingly necessary as the industry became more technical. In 1834, the Mechanic's Institute was established and held lectures in Alexandria, on the Leven, about science and literature for the benefit of local workers.[77]

Individual accounts from workers in the Vale of Leven in the late nineteenth century provide some insight into daily life in the Turkey red industry. Work was physically demanding, and many suffered strain injuries from carrying heavy fabric. A local doctor reported in 1843 that liver problems were common from an early age and many physical injuries were documented.[78] Chronic ill health was rampant in the textile industry. A source describing dyeing in Darnétal in the late eighteenth century says that methods that used quicklime "made holes in the legs of the workers" who were using their feet to agitate the cloth.[79] An 1849 article in *Scientific American* noted that "[M]any people have lost their lives while engaged at the occupation of discharging Adrianople red handkerchiefs" due to the toxic chlorine gas released when the bleach and acid react. Strain injuries from moving heavy goods, burns and exhaustion from heat exposure, injury from high-speed machinery with little to no safety precautions, inhalation of dust and lint, and exposure to chemicals were all regular hazards in textile spinning, weaving, dyeing, and printing. John Jardine, a block printer at Dalmonach, told an inspector that he had a seven-year apprenticeship and began work at age six as a tearer, wetting blocks for adult textile printers. At the time he was working from 6:00 a.m. to 6: p.m., but previously had worked from first light with a later start in the winter. He was concerned that his children, who worked more than he did at their age, were always tired and unable to attend night school. Beating children in the factories had become less common, but they were still the victims of accidents from the unsafe working conditions. Older boys like John Mitchell and James Newlands, aged thirteen, who could do heavier labor but were not skilled enough for more advanced positions, worked alongside women washing cloth in the river, a cold, wet, and heavy occupation. They worked from 5:00 a.m. to 8:00 or 9:00 p.m. with forty-five minutes for meal breaks. Mary Brannen worked from 6:00 a.m. to 9:00 p.m. carrying wet fabric from the wringers to the steaming equipment. Artificial lighting enabled longer operating hours, so she sometimes worked night shifts as well. In the bleaching department, Janet Miller received thick aprons from her employer for protection but had to supply her own leg coverings and shoes. Mary Moody and Mary Maxwell worked in the stoves, sometimes fainting from the heat. A fifteen-year-old named Martha McKechnie worked in the finishing house folding cloth from 6:00 a.m. to 7:00 p.m., which prevented her from attending night school. She and many of her colleagues were illiterate, and it was common for children to only go to school from ages five to eight.

In the mid-nineteenth century, an inspector noted that a bleacher earned eleven shillings per week (10 shillings is about £40 today) and lived with his wife and four children in a room about 9 x 18 feet (2.7 x 5.5 m) and 7 feet (2 m) high, with furniture and a small garden. They had porridge for breakfast and supper, with a pint of milk. For dinner they had a broth with a small piece of meat or fish, oat cakes, and potatoes. Wages for a machine printer were 30–50 shillings weekly, 25–30 shillings for small block printers, 30–40 shillings for large block printers, and 4–7 shillings for boys.[80] Records from 1872 for the Hamilton works in Massachusetts show low wages and layoffs were common. Most workers earned $1–2 per day, while another who was possibly a child earned sixty cents ($1 in 1872 is approximately $21 today).[81] Historical data for housing costs indicates a four-room home in Boston in 1872 could be rented for $10 monthly, so it took about ten days' work at $1 per day to make rent.

Employers exerted a fair amount of control over life, as shown in some accounts from the Vale of Leven. When fabric was dried outdoors, it was spread in fields outside the dyeing buildings and proved to be a tempting target for some thieves. Newspapers from the late eighteenth and early nineteenth centuries reported various incidents. For example, Catherine Veer stole printed cotton shawls from the Littlemill bleachfield near Dumbarton, for which she was banished from the trade for fourteen years. Sometimes penalties even included being sent to royal plantations in the Caribbean. By 1898, 7,000 people were employed in the Vale of Leven producing Turkey red textiles.[82] Competition was fierce among local firms due to the similarity of the product and the practice of stealing designs through copyright infringement and industrial espionage. Workers were often subject to what would be called non-compete clauses today, and it was not unheard of for firms to threaten the employment of entire families if one member indicated they wanted to leave.[83]

Colonialism

As discussed in Chapter 2, Turkey red dyeing likely originated in India. The eastward expansion of European trade and the drive to obtain knowledge of production techniques for the goods they imported, along with the mechanization of the Industrial Revolution and the expansion of colonial empires, were ideal circumstances for industrialized nations to undermine domestic production of goods in countries for whom they had once been customers. British and French textile manufacturers could not initially compete in the markets of the Levant and India, where local labor was cheap and the skill level much higher. This changed with the development of the spinning jenny, water frame, and spinning mule, which enabled the large-scale production of fine yarns at an affordable price. Now, it was commercially viable to sell cheaper European textiles in India and the Ottoman Empire. Cotton yarn was the first product to be exported from England to the Levant starting in the 1790s. In 1814, it was observed that cheap English cotton fabrics were in high demand in markets from Cairo to Istanbul. Hand manufacturing could not match the scale of machine production, and was gradually displaced by foreign goods. Competition had a significant effect on local manufacturing, pushing it into decline by the mid-1820s and by the 1860s many trades had disappeared.[84] This new industry demanded more raw cotton than had ever been previously grown, causing cotton to shift from being a subsidiary crop for farmers growing foodstuffs to the large-scale, dedicated cotton plantations located in the southern United States.[85] More production meant more enslaved labor was sought, and Native people were forcibly relocated from land intended for new cotton plantations.

The Indian market was flooded with cheap, machine-made English prints that imitated local styles.[86] It took hand-spinners in India 50,000 hours to process one hundred pounds of raw cotton, which could be accomplished by a spinning mule in a fraction of the time. By the early nineteenth century British yarn cost one-third of what Indian yarn did, and it was cheaper for Indian weavers to work with imported product to make their fabric.[87] An 1840 article in *The Saturday Magazine* discusses bandana handkerchiefs, an increasingly popular accessory. They glibly describe the fifty years of rising market competition between Britain and India as "a rivalry of a remarkable kind."[88] More accurately, British industry was steadily eroding the market for Indian textiles in both countries. By the end of the nineteenth century, Britain had caused India to shift from exporting their own goods to importing textile materials through their industrial and political power.[89] British interests in India worked to block supplies of goods from other European nations and repress local industry that would otherwise compete with British exports. The intent was to force Indian textile manufacturing into extinction, making the country entirely dependent on British imports. Although successful on an economic scale, the campaign could not completely quash the production of local handmade goods. Some of these makers were working on a hybrid system, weaving with imported British yarn, or printing on imported cloth. Their familiarity with local taste made it possible to compete on a smaller scale.[90]

In the late nineteenth century, efforts were made to establish technology-based industry in India by studying practice in Britain and America. Two Indians, Keshav Bhat and Bhaskar Rajwade, traveled abroad to learn about Western manufacturing practice. Bhat traveled first to Britain, but despite the letters of introduction he carried from Mumbai merchants he could not gain access to any firms because he was seen as "a spy prying into the secrets of their industries." In 1882, he went to America to study at the Massachusetts Institute of Technology and attempted to work in the dyeing trade there. Through a series of introductions starting in India, Bhat was able to gain access to the Cochrane works in Malden and learn about Turkey red. He returned to Poona (now Pune) and opened the Indo-American Dye House Company, though it was not particularly successful.[91]

It is ironic that European knowledge of Turkey red dyeing came from the migration of skilled workers and various episodes of industrial espionage considering that a century and a half later, a citizen of the land from which Turkey red practice likely originated was thwarted in his efforts to gain knowledge of current industrial practice and was unsuccessful in his attempt to establish manufacturing back home.

Conclusion

Turkey red dyeing disseminated westward into Europe from the Levant, as discussed in Chapter 2, with activity increasing in the early eighteenth century. By the mid-eighteenth century, when practical knowledge of Turkey red reached France and Britain, the textile industries there were on the verge of adopting the new machinery of the Industrial Revolution. This allowed production to outpace hand manufacturing in India and the Levant, eroding the markets where European traders once purchased the goods they were now imitating. The abundance of cheaper, machine-made textiles enabled more European consumers to purchase items that would previously have been unaffordable. Advertising as we know it today also developed during this time via efforts to attract spending by these new consumers. Popular taste and mechanical reproduction changed the design process, prioritizing speed and removing artists further from production. Labor also changed, and factory workers in textile dyeing and printing had hazardous, physically demanding jobs with few protections. The major Turkey red dyeing and printing locations during the nineteenth century were England, Scotland, France, Switzerland, and the Netherlands. This chapter summarizes some of the firms and locations associated with Turkey red production but does not claim to be a comprehensive study, particularly for the industries in Rouen, Mulhouse, and Switzerland for which there are not yet any histories focusing on Turkey red.

Trade, Use, and Object Record

As discussed in Chapter 1, the history of textile production and consumption can be difficult to establish. Our ability to understand how, where, and by whom textiles were used is contingent upon available documentary evidence and collection objects, which may or may not present an accurate record based on what is preserved. This is especially true for quotidian materials, which were less prestigious and more likely to be reworked into new items like quilts or rags and used until no longer serviceable than they were to be preserved. Additionally, prior to industrialization textiles were labor-intensive handmade products, so there are also relatively fewer surviving articles from before the eighteenth century which limits interpretation of availability and use.

Taste and availability also influenced what was created, and therefore what has been preserved. Turkey red was popular in places where red was a propitious color, and in warmer climates like India and parts of China where cotton was preferred over wool for comfort. Although the European industry was well-established by the mid-nineteenth century, it was not often seen there in everyday garments outside of those worn by workers in the industry, who would have had access to discounted or slightly imperfect wares. One contemporary commentary declared wearing red was "evidence of immaturity for women in the fall time of life." It was also said to be more suitable "among savage races" and to be associated with a lower degree of civilization.[1] The Great Exhibition of 1851 in London (see Figure 5.5) and the Glasgow International Exhibition in 1888 expanded the domestic audience. Commentary from an account of the 1888 exhibition nevertheless said that the textiles had a "peculiar pattern and look ... not intended for the home market," and were considered to be more suitable for Australia, New Zealand, Brazil, and West Africa.[2] In her autobiographical novel *Little House in the Big Woods*, American writer Laura Ingalls Wilder recounts a family trip to the general store from their farm in Wisconsin, sometime in the early 1870s. Her father encouraged her mother to choose fabric for a new apron. She protested, and he told her she must pick something or "he would get her the turkey red piece with the big yellow pattern," prompting her to select a demure pattern of rosebuds on a tan ground.[3] Another source, from 1880, observed that "turkey red is revived, in the form of turkey-red calico, as worn forty years ago."[4] Regardless of the seemingly mixed opinions, the evidence discussed in this chapter indicates Turkey red was available and used to make a myriad of useful and decorative textiles around the world throughout the long nineteenth century, the period from roughly 1789–1914.

Historical newspapers are a valuable resource for tracking the availability of Turkey red, especially with digitized, searchable archives available online. Preservation rates for original publications vary since they were typically printed cheaply on paper not intended to last, and decades of daily or weekly editions quickly makes for a large archive that can be costly to digitize. The examples discussed here, which only include English-language publications, are not presented as a complete study but rather to illustrate its popularity and availability. The purpose of many of these announcements, which would have been paid for by word or column length, was to let consumers know a product was available. Turkey red is named, but never described, implying the general population needed no further explanation of its properties or reputation.

Historical imagery, like paintings and photographs, also provides some visual evidence for Turkey red use. Identifying fiber types from an image cannot always be conclusive, but may be inferred from historical and geographical context, the drape, who the sitter is, and whether there is any pattern or design provide additional clues. For example, a late-sixteenth century watercolor drawing of a Turkish woman by the English artist John White in the British Museum collection depicts her wearing a bright red overcoat with a wide, flared skirt (cat. no. 1906,0509.1.32). Wool is a practical material for outerwear and was easy to dye red with natural dyes, though the cut of the coat implies it could also be made with a woven cotton, possibly Turkey red. In this case, a stronger conclusion cannot be made without more information. Colorized photographs like Figure 2.5 can be visual records, provided (as with paintings) that the artist accurately depicted the true colors. Autochrome images, an early form of color photography, capture information that cannot be as reliably determined from black and white images. The *Archives de la planète,* a digital database of autochrome images from the Albert Kahn collection, contains possible evidence of Turkey red use in the early twentieth century from around the world. For example, a 1913 autochrome from Fez, Morocco shows a woman carrying her baby in a sling and wearing a bright red headscarf with a white print (cat. no. A832S). In a warm, sunny climate such an object is likely to be cotton, and therefore Turkey red. Likewise for the turban and sash worn by an Indian man in cat. no. A4364.[5]

This chapter explores how Turkey red was used, and by whom, through a selection of representative and distinctive objects, publications, archive records, and artworks. The discussion here is not intended to be a thorough investigation, but rather a broad overview of Turkey red in a variety of cultures from the late eighteenth to early twentieth centuries. Like many of the objects discussed in this book, no analysis has been done to confirm their identity, but it is proposed to be Turkey red based on the criteria discussed in Chapter 1.

Documentary Evidence of Availability

Turkey red appears by name in advertisements and cargo manifests from at least the late-eighteenth century onward. The context, and target market of these publications indicate more about who consumed Turkey red. Early records document Turkey red trade from the Levant and the beginning of dyeing in Britain. On March 6, 1759, the *Manchester Mercury* advertised an auction that included two lots of Turkey red cotton yarn and one of red cotton twill, all dyed in Turkey.[6] In 1778, a notice in the *Dublin Journal* said that Captain Roach of Dublin and the schooner *The Rambler* had captured *Le Victoire* near Livorno while she was traveling from Smyrna to Marseille, carrying among other goods eighty-seven bales of fine Turkey red cotton yarn and sixty-seven bales of madder root.[7] By 1799, an advertisement in the same paper said that Alex Ramsay and Company from Glasgow were selling Turkey red yarn, pullicate handkerchiefs, and muslins of various descriptions. The announcement went on to say that Ramsay "was not inclined to sound his own praise, but … begs to observe a fact known to every Manufacturer in Scotland … that his Turkey red Dye was preferred there to that of England."[8] A 1758 description of a working-woman's wardrobe in Britain includes a "calico red and white bedgown," and a 1789 plan to provide essential clothing for the working poor recommended young boys receive "red napped waistcoats," likely made of fustian or cotton cloth.[9] While not named specifically, Turkey red was a durable and attractive option for working-class people, and it is known that Turkey red yarn was imported from the Levant at this time. Although Turkey red was a premium textile product in terms of quality and more expensive than plain cloth, it was not high-status or particularly expensive compared to fine silks and woolens and the growth of the industry over the nineteenth century indicates it was within the means of the average consumer.

Turkey red textiles were routinely sold on the North American market as well based on the activity of Scottish mercantile firms. An advertisement for Scottish Turkey red bandanas sold by Tappan and Searle of Boston, Massachusetts, appears in the *Providence Gazette* on October 10, 1812.[10] In the early nineteenth century, railroad infrastructure in North America was limited to the more densely populated northeast. West of the Appalachian Mountains, goods were typically transported by steamboat, often up the Mississippi from the port of New Orleans and then via smaller waterways or overland to their destination. Some limited rail lines were built in the south to transport cotton, but there was no interconnected network before the late 1860s so goods often moved slowly. A July 1803 notice in the *Kentucky Gazette and General Advertiser* of Lexington, Kentucky says that John Jordan had Turkey red merchandise for sale.[11] In Hagerstown, Maryland, an advertisement from 1806 said that G.M. Conradt had taken over the dye business of Daniel Nead and that he stocked Turkey red cotton in addition to his own products.[12] The *Savannah Republican* in Georgia carried an 1816 advertisement from Andrew Low & Company for the "First Fall Goods" arrived on the ship *Lucy* from Liverpool, which included Turkey red yarn.[13] The Tuscaloosa, Alabama *State Intelligencer* advertised Turkey red prints recently arrived aboard the steamboat *Catawba* in 1831.[14] The Tarboro, North Carolina *Free Press* printed an 1833 advertisement from James Weddell for new spring and summer goods, including Turkey red cotton.[15] That same year, J.W. Armstrong & Company advertised in the *Kingston Upper Canada Herald* that they received from Liverpool and Glasgow an assortment of goods including Turkey red.[16] The *Boston Morning Post* printed an advertisement in 1840 that J.B. Prince on Milk Street had Turkey red cotton for sale.[17] Some Turkey red advertisements appear on the same page as those for cheap garments and shoes to clothe enslaved people, or with notices for individuals for sale, lease, or capture (see Figure 6.1). It is important to acknowledge that an integral part of the history of cotton textiles, including Turkey red, is the dependency on enslaved labor, and the stark juxtaposition of advertisements for human beings and for the products of their forced labor is an effective reminder.

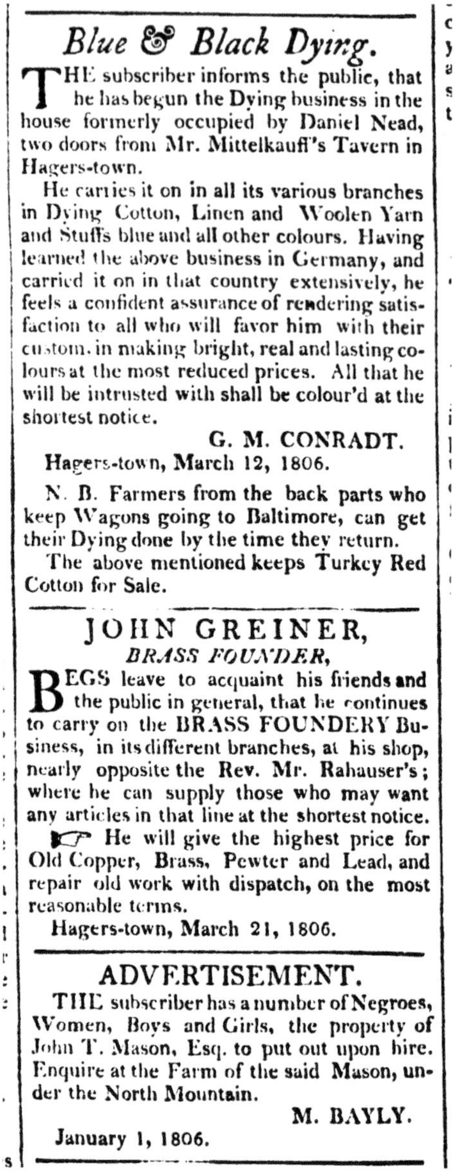

Figure 6.1 Advertisements in the Hagerstown, Maryland newspaper for Turkey red goods and for enslaved labor, which was part of the Turkey red supply chain. *The Hagerstown Maryland Herald and Hagerstown Weekly Advertiser. May 9, 1806*

By the mid-nineteenth century, advertisements appeared in publications in what was the western frontier of the United States. In 1850, the *Rising Sun Indiana Whig* printed an advertisement from W.P. White & Company for new goods, including "Turkey red and embroidered curtain muslin."[18] A 1879 advertisement in the *Columbus Journal* in eastern Nebraska listed Turkey red handkerchiefs for $0.05 at the New York Cheap Cash Store.[19] In Lincoln, the state capital, the *Lincoln Daily Evening News* printed an advertisement in 1883 for Turkey red table damasks and imported Turkey red table and stand covers for sale at the Osborne, Kespohl & Co. department store.[20] The *Goshen Weekly News* in Indiana promoted Turkey red damask for sale in 1895, ranging from $0.15 to $0.23.[21] On the western side of the continent, Turkey red was advertised in an 1849 edition of *The Californian*, which carried a notice that Dring's stores in Adobie and Fort Sutter had recently received goods including Turkey red prints imported by various ships.[22] The California Gold Rush had begun in early 1848 with the discovery of gold in the northern part of the state. At the time, California was a very remote territory relative to the rest of the United States, which had recently acquired it. Before the Transcontinental Railroad in the mid-nineteenth century and the Panama Canal in the early twentieth century, travel from East Coast required sailing around the southern tip of South America, a treacherous overland journey by wagon or horseback across the plains and the Rocky Mountains, or sailing to Panama or Mexico and making an overland crossing before boarding a northbound ship. Prospectors arrived in waves from around the world, with the first large group of Americans from the East Coast arriving in 1849, earning them the nickname "forty-niners." Supplies for the rapidly growing population were in high demand, and the early presence of Turkey red arriving on shipments to California indicates its position as a desirable product. It was shipped even further west to the Hawaiian Islands, which from 1795 to 1893 were an independent state known as the Kingdom of Hawai'i. An 1855 advertisement in *The Polynesian* for B.F. Snow mentioned goods recently arrived from Boston on the Cato, including Turkey red chintz.[23] The *Honolulu Pacific Commercial Advertiser* in 1860 carried a notice from Janion, Green & Co. for goods recently arrived on the *Humphrey Nelson* from Liverpool, which had sailed about eight months prior, and had Turkey red cloth and red and yellow prints (see Figure 6.2).[24] Shipments of goods to the western Pacific Ocean also included Turkey red. The *Hobart Town Gazette and Van Diemen's Land Advertiser*, on the island now called Tasmania, printed an advertisement in 1822 for goods recently arrived, including Turkey red cotton and bandana handkerchiefs.[25] An 1826 edition of the *Sydney Monitor* said that T.G. Pitman was selling Turkey red handkerchiefs at his premises on George Street.[26] In 1920, *The Japan Advertiser* said Nippon Kokusan Co. Ltd. sold Turkey red shirting.[27]

How Turkey Red was Used

Turkey red yarn and fabric, both printed and plain, served a variety of purposes in decorative and utilitarian textile objects. With the exception of bandanas or handkerchiefs, there is little evidence of a single popular item closely associated with Turkey red, though it is generally associated with objects that were more likely to be heavily used and washed. It is known from

Figure 6.2 Advertisement in the Honolulu Pacific and Commercial Advertiser for imported Turkey red from Britain. *The Honolulu Pacific Commercial Advertiser. January 5, 1860*

written records that Turkey red printed saris were produced and exported in the nineteenth century, but this research did not identify any in collections.[28] This illustrates the difficulty in obtaining an accurate impression of Turkey red use and the potential scope of unidentified Turkey red in collections. The objects and artworks discussed here represent some of the common and uncommon ways it was used.

Bandanas

Turkey red bandanas had global appeal during the nineteenth and early twentieth centuries and were made in a vast array of designs and colorways. They functioned as headcovers, neckwear, slings to carry items, and served as an eye-catching, affordable, and durable way to commemorate events and disseminate information. A listing in the 1875 spring Montgomery Ward catalog for a dozen men's unhemmed Turkey red handkerchiefs (smaller than a bandana) was $2 (about $45 today), comparatively a dozen hemmed regular handkerchiefs was $1.[29] Bandanas were a major Scottish export product in particular, but also made by Swiss, Dutch, and American firms, and were sold in Europe, Africa, Asia, the Caribbean, and the Americas.[30] Although most have not been analysed to confirm they are Turkey red, there is no evidence in the literature to suggest red bandanas were made from anything else.

Discharge-printed bandanas were first made on a large scale in the early nineteenth century (see Chapter 4). They often had a pattern of white polka dots on a red ground in imitation of tie-dyed Indian silk kerchiefs, a style that became the classic Turkey red bandana as shown in Figure 6.3.[31] The young woman depicted in the 1849 painting Peasant Girl with a Scarf by French artist Gustave Courbet in Figure 6.4 appears to be wearing such a bandana around her neck, and another as a

Figure 6.3 Turkey red bandana *c.* 1870–6 printed by the Archibald Orr Ewing firm of Alexandria, Scotland. *Glasgow Museums and Libraries Collections. Gift of Archibald Orr Ewing Company. 1876.128.b.15*

Figure 6.4 The young woman in Peasant Girl with a Scarf, painted *c.* 1849 by Gustave Courbet, wears a red and white scarf with a border characteristic of Turkey red bandanas. *Norton Simon Art Foundation. M.1989.2.P*

head covering. The National Trust Collections hold a Turkey red bandana in this style (cat. no. 642106) originally owned by the children's author and illustrator Beatrix Potter. Potter used it as a model for some of her work, notably when Peter Rabbit loses his clothes in Mr. McGregor's garden in *The Tale of Benjamin Bunny*, shown in Figure 6.5.

More elaborate patterns, multi-colored prints, and commemorative designs were also made. The latter, typically with only black and discharged white on red, could be designed and printed fairly quickly, which was useful for marking to current events. A volume in the Board of Trade Representations and Registers of Designs at The National Archives in London contains a pattern from March 1885 for a printed Turkey red handkerchief featuring General Charles George Gordon, "Gordon of Khartoum," a British army officer who had fought in the Crimean War and the Taiping Rebellion and had died on January 26, 1885. Many political events were commemorated, like the

Figure 6.5 A watercolor illustration from *The Tale of Benjamin Bunny* depicting Peter Rabbit wrapped in a red and white printed pocket handkerchief since he lost his clothes in Mr. McGregor's garden. *Illustration from The Tale of Benjamin Bunny by Beatrix Potter © Frederick Warne & Co. Ltd., 1904, 2002*

1898 ascent of Dutch queen Wilhelmina in Figure 6.6. The National Kerchief Company printed commemorative bandanas as seen in Figure 6.7 for the 1912 presidential campaign of Teddy Roosevelt, who had served as President from 1901 to 1909 for the Republican party and was running again for the Progressive "Bull Moose" Party. It is unclear whether there was any intended parallel between "TR" for Roosevelt's initials and Turkey red, or whether the manufacturer dyed the fabric or just printed bandanas.

Commemorative handkerchief motifs in the National Museums Scotland collection include the end of the Crimean War in the 1850s and Queen Victoria's 1887 Golden Jubilee.[32] Even the 1885

Figure 6.6 Printed handkerchief commemorating the ascension of Queen Wilhelmina of the Netherlands in 1898. *Rijksmuseum. Herinneringszakdoek met troonsbestijging Koningin Wilhelmina in 1898. Gift of E.T. Scheltinga Koopman, Nigtevecht. BK-1974-23*

invention of the rabies vaccine was commemorated with a French Turkey red bandana honoring Louis Pasteur and Émile Roux, seen in Figure 6.8. Other examples include the four-paneled design shown in Figure 6.9 depicting a lacrosse game, lawn tennis, and some classic children's rhyming songs, likely produced in America where lacrosse originated. An 1895 example in the Kalmar Läns Museum collection (cat. no KLM032776-00001), called an "instruction sheet," has diagrams and directions in Swedish for how to butcher oxen. In the American Museum of Natural History collection, there is a red bandana with a pattern of flowers and cloverleafs in white, blue, and black

Figure 6.7 A 1912 campaign bandana for Teddy Roosevelt, who had served as President from 1901–9 for the Republican party and was running again for the Progressive "Bull Moose" Party. It is unclear whether there was any intended parallel between "TR" for Roosevelt's initials and Turkey red. *Courtesy of the Theodore Roosevelt Inaugural Site Foundation—Buffalo, NY. 1982.003*

from the turn of the twentieth century (cat. no. 40.1/5954) that was acquired in Colombia, and is likely European Turkey red.

From the second half of the nineteenth century, Turkey red took on associations with the frontier and the image of the cowboy in American cultural folklore. The expansion of cattle ranching in the 1880s was financed by British investors, particularly Scottish ones. They would send their own managers to oversee the American staff, bringing with them the habit of using Turkey red bandanas.[33] Frontier associations are also illustrated in a short story called "Turkey Red" by

Figure 6.8 Printed handkerchief commemorating the 1885 invention of the rabies vaccine by Louis Pasteur and Émile Roux, late nineteenth-century France. *Cooper Hewitt, Smithsonian Design Museum. Gift of Mr. and Mrs. Simeon Braguin. 1981-11-18*

American author Frances Gilchrist Wood, which appeared in *The Best Short Stories of 1920* and was published in 1935 as a full-length book. It depicts the hardships faced by settlers on the plains following the 1862 Homestead Act, including droughts, severe blizzards, crop failures, pests and wildlife, conflicts with Native People, illness, and isolation. The protagonists take refuge from a blizzard at a homestead where a young wife is alone holding the family claim and caring for a severely ill baby while her husband works in a larger town over the winter. A typical homesteader's claim shanty was a quickly-built structure meant to be replaced with a more permanent home when the resources were available. Inside the shanty, a single room of eight by ten feet, was a bed, a stove, and a bench. Although the unfinished boards of the walls and roof were covered with old newspapers, "cushions and curtains of turkey-red calico brightened the homely shack." Hillas, one of the travelers, muses over the red cactus flowers that bloom on the prairie during extreme summer

Figure 6.9 A printed handkerchief, late nineteenth century. The four panels depict activities and rhymes for children in black and white with red accents and borders. *Philadelphia Museum of Art. Gift of the Philadelphia Commercial Museum (also known as the Philadelphia Civic Center Museum), Philadelphia, Pennsylvania, 2004. 2004-111-51*

droughts and the parcel of paper-wrapped Turkey red in his pocket. He says, "up and down the frontier in these shacks, homes, you'll find things made of turkey-red calico, cheap, common elsewhere … it's our 'colors.'"[34] Here, the Turkey red is symbolic of the resilience of the frontier settlers, the otherwise ordinary becoming extraordinary in a place of hardship and shortage, and the need for a bit of brightness in a difficult existence.

Domestic Textiles, Quilts and Bedcovers

The wash fastness and resistance to bleaching of Turkey red made it particularly useful as a decorative element in domestic textiles, which were subject to more frequent and vigorous laundering. One common use was to weave yarn into white textiles like the serviette in Figure 6.10.

Figure 6.10 A cotton finger towel woven with Turkey red and white yarn. The fabric was bleached to finish, and the durability of the Turkey red meant it could withstand regular washing and bleaching without fading. *Rijksmusem. Vingerdoek van katoenen damast met rozenpatroon en afgezet met rode rand. Gift of M. van Gogh, The Hague. BK-1973-200-K*

Irish linen finishers purchased cotton Turkey red yarn from Scottish dyers for weaving decorative patterns into their wares, and these items seem to have been popular for some decades.[35]

A swatch of a towel in a white pattern on a red ground shown in Figure 6.11 was made in Austria in the 1820s, while Figure 6.12 shows a 1907 autochrome image of two men playing chess on a similarly-patterned table covering. Many sizes and qualities of Turkey red table covers are found in catalogs for domestic goods.[36] An 1875 American retail catalog lists a dozen Turkey Red Napkins with "neat patterns" for 92 ½ cents (about $23 today). Comparatively, the same number of plain bleached napkins is 85 cents. Turkey Red Table Damask was around 90 cents per yard, while regular Table Damask was 27 ½ cents per yard (about $7).

In the late nineteenth and early twentieth centuries, cotton embroidery thread was sold in red, blue, and black. The blue was most likely indigo, while the red would have been Turkey red (see Figure 6.13), which was marketed based on its exceptional fastness.[37] Samplers, like the one shown in Figure 6.14, were demonstrations of stitching exercises often done in high-contrast colors like red on a white ground. It was also possible to decorate one's own white linens or decorative objects like the embroidered panel in Figure 2.7. Homes were also decorated with Turkey red upholstery, typically bearing floral designs that Scottish firms made for the British and European markets. Furnishing fabrics in the "two red" style designs (see Figure 4.6) were especially popular. There is

Figure 6.11 Sample of a coffee towel woven by master weaver Anton Langer from linen and Turkey red cotton. Made in Pottenstein, Austria in 1821. *MAK—Museum of Applied Arts. TGM 30455*

enough evidence from newspaper advertisements for drapers and furniture retailers to indicate that Turkey red was a regular feature in the decoration of Victorian and Edwardian homes.[38] No doubt the solid, red hue was appealing in both daytime and evening lighting, and its resistance to fading meant it would look nice for longer than other options might.

Plain and printed Turkey red appears in many quilts from the nineteenth and early twentieth centuries, which becomes apparent once one has an eye for the characteristic palette of the prints. Quilts could be made with new or repurposed fabric, and durable Turkey red scrap and rags were often used . An example of an "Ohio star" block pieced with printed Turkey red is shown in Figure

Figure 6.12 Autochrome photograph by Alfred Stieglitz taken in June 1907. The image shows two men playing chess at a table covered with a cloth woven from Turkey red and white yarn. *Metropolitan Museum of Art. Gilman Collection, Purchase, Mr. and Mrs. Henry R. Kravis Gift, 2005. 2005.100.476*

6.15. An appliqué tulip, where the design piece is stitched onto the top of a larger piece of fabric, is shown in Figure 6.16. Turkey red is often identified in quilts by its distinctive streaky wearing pattern, resulting from the color complex deposited on the surface of the fiber wearing away. Some examples of quilts are discussed here, and more thorough discussions of quilt history can be found in *Clues in the Calico* by Barbara Brackman and *Down by the Old Mill Stream: Quilts in Rhode Island* (eds. Welters and Ordoñez).[39]

Turkey red and white was a popular color scheme for quilts for some decades beginning in the 1840s, and many North American quilts made from 1840 to 1870 incorporated both plain and printed red into the block designs. An early twentieth-century example with plain red is shown in Figure 6.17. In the nineteenth century, appliqué quilts made in red and green on a white background, sometimes complemented with yellow or pink, were also popular (see Figure 6.18). The popularity may be due to the ease of depicting natural motifs, e.g., red flowers with green foliage, and according

Figure 6.13 Spools of Turkey red embroidery thread from the Dexter Yarn Company, early twentieth-century American. The right spool says it is "boil proof" and the box (not pictured) says "guaranteed fast." *Julie Wertz*

to one source women in Appalachia used these colors because they were more reliable. This implies the red fabric was Turkey red, and also indicates that the green and yellow were chrome yellow and Prussian blue if the fastness was comparable to that of the red. The style went out of fashion as more synthetic dyes came to market, though early synthetic reds (like the product in Figure 1.2) faded to tan or brown. The outbreak of the First World War affected the availability of textile goods and contributed to the decline of Turkey red use in American quilting. In addition to difficulties importing European fabric to the United States, the war and trade embargoes interrupted the supply of synthetic German dyes, including alizarin, to British and American textile manufacturers.[40]

Turkey red was also used in other, non-quilted bedcovers. An early nineteenth-century example of a French bedcovering, in the characteristic palette of green, yellow, black, blue, and white on a red ground, is likely printed Turkey red (see Figure 6.19). Traditional textile practice in Hawaii also incorporated imported Turkey red into barkcloth or *kapa*, which was made by beating plant fibers

Figure 6.14 A sampler stitched in red cotton on a white cotton ground by Louise Wilhelmina Nulle around 1900. The contrast of the Turkey red embroidery thread was striking and was not likely to fade or bleed onto the white. *Rijksmuseum. Merklap van rode katoen op katoen. Schenking van de Vereniging Het Kantsalet. BK-1975-389*

from the paper mulberry (*Broussonetia papyrifera*) into a thin, supple non-woven fabric. Since it could not be immersion dyed, color was typically a surface application painted onto the textile. Hawaiian women adopted appliqué techniques of decoration in their *kapa*. To incorporate red in the later nineteenth century, small pieces of imported Turkey red were shredded into fibers and spread over the bark fibers before beating. The process, called *pa'i 'ula*, means "beaten in red" and made a fabric with a mottled red color. This red fabric could be further cut and used as an appliqué to decorate uncolored *kapa*, much like the appliqué quilt in Figure 6.16. One of the earliest surviving examples of Turkey red appliqué *kapa*, held in a private collection, was made between 1839 and

Figure 6.15 A quilt block in an Ohio Star pattern made with printed Turkey red, *c.* 1840–60. *Courtesy of Barbara Brackman*

1858 by Ka ʻaia Kuawalu. The *paʻi ʻula* were often used as bedcovers. An example from the 1866 expedition of Prince Alfred to Hawaiʻi, is shown in Figure 6.20. Embedded fragments of Turkey red cotton give it a pinkish cast when viewed from arm's length.[41]

Accessories, Garments, and Tools

Turkey red appears in a plethora of finished clothing and accessories. A shipment to Grenada in May 1791 contained textiles valued at £1,649 (roughly £125,000 today), including items like striped and checked cottons and handkerchiefs made with white linen and Turkey red yarn imported from

Figure 6.16 Quilt block featuring an appliqué tulip design made of printed Turkey red, *c.* 1 840–60. *Courtesy of Barbara Brackman*

the Levant. Similar garments are shown in the painting in Figure 1.6, where women wear white shawls with red borders and designs consistent with descriptions of pullicates woven in Scotland during that era.[42] In the sunny tropical climate of the Caribbean, Turkey red would be a visually appealing color and robust to continued light exposure. These early Turkey red exports were often intended for use by enslaved people and people of color during this period.[43] A portrait of a young woman of color painted in Louisiana around 1840 by Jacques Guillaume Lucien Amans shows her wearing a bright red tignon (see Figure 6.21). The hue, the drape of the fabric, the warm climate, and the history of sumptuary laws regarding head coverings for Afro-Creole women all indicate it is likely Turkey red.[44]

Quilted petticoats and housecoats made of printed Turkey red were a popular off-the-rack product in late nineteenth-century Britain, and were being made at home as early as 1842. The fabric used in these garments was usually a plain weave. An example of a late nineteenth-century petticoat made by the Booth and Fox firm is shown in Figure 6.22. Another company, McLintock and Sons of Barnsley, used carded silk waste, or noils, to make a lightweight, warm batting about

Figure 6.17 An early twentieth-century American quilt Steeplechase quilt in Turkey red and white fabric. *Courtesy of Barbara Brackman*

Figure 6.18 An American appliqué quilt *c.* 1850 depicting flowers, birds, hearts, and cats in red and green on a white ground. As well as the contrasting colors being visually appealing, the Turkey red and chrome green fabric had good fastness to light and washing. *Metropolitan Museum of Art. Gift of Allison L. Reddington, in celebration of the Museum's 150th anniversary. 2020.69*

Figure 6.19 An early nineteenth-century French bedcover in a colorway characteristic of printed Turkey red. The pattern is more painterly than many of the geometric designs preserved in pattern books and was made near Rouen. *Philadelphia Museum of Art. Printed Textile: "Les Heureux Epoux". Made by Bouvet fauquier à Bolbec. Purchased with the Art in Industry Fund from the Henri Clouzot Collection, 1937. 1937-11-128*

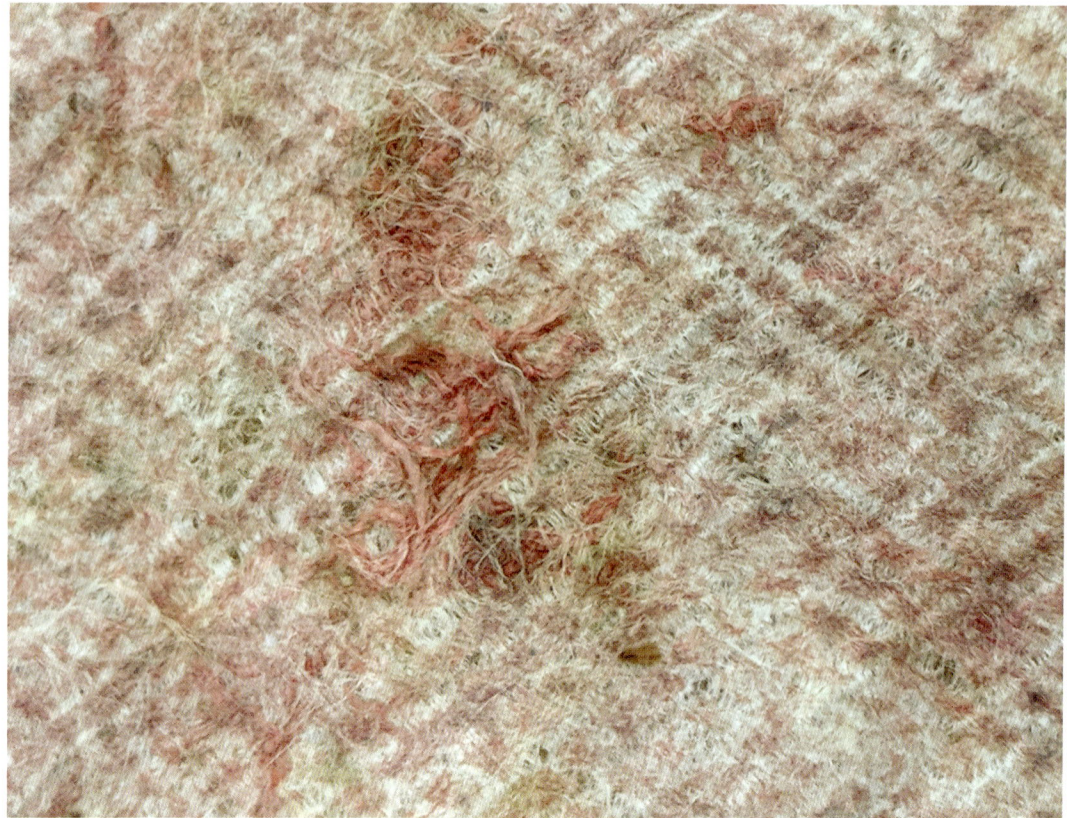

Figure 6.20 Turkey red fibers beaten into a *kapa* to make a *pa'i'ula* given to Prince Alfred, Duke of Saxe-Coburg and Gotha in Hawaii in 1866 on his world tour. *Glasgow Life. 1879.42.hn.2*

one yard wide, which they called "Toralium." This fit nicely between two pieces of Turkey red fabric, which were quilted together. Later, though it is not clear when, the firm switched to cotton down. Women's fashion in the mid-nineteenth century was built around the full crinoline, a stiff structural underskirt that became increasingly outsized. After fashionable silhouettes became slenderer and did not require padded skirts, McLintock's continued to make and re-cover down-filled quilts.[45] They also made quilted housecoats like the one in the National Museums Scotland collection (cat. no. A.1983.128), a warm and cheerful option for domestic wear.[46]

Its bright hue and durability made Turkey red ideal for children's garments, like the dress in Figure 6.23 and the coat in Figure 6.24. The British Museum collection has a late nineteenth-century child's jacket from India in red cotton with a print characteristic of Turkey red (cat. no. As1933,0715.318). This also applied similarly to swimwear, like the bathing costume in Figure 6.24, which would be repeatedly exposed to water and sunlight. A similar garment is worn by the sitter Christina Bevan in the 1913 autochrome image *Christina* by Mervyn O'Gorman (Royal Photographic Society Collection (cat. no. RPS.622-2020)). It also served well for aprons, which needed more washing than the garments they covered. In an autochrome image from France

Figure 6.21 Portrait of a young woman of color wearing a bright red tignon. The warm climate of New Orleans make it likely she was wearing cotton, and therefore Turkey red. T*he Historic New Orleans Collection. Creole in a red headdress. Jacques Guillaume Lucien Amans, c. 1840. Acquisition made possible by The Diana Helis Henry Fund of The Helis Foundation in memory of Charles A. Snyder. 2010.0306*

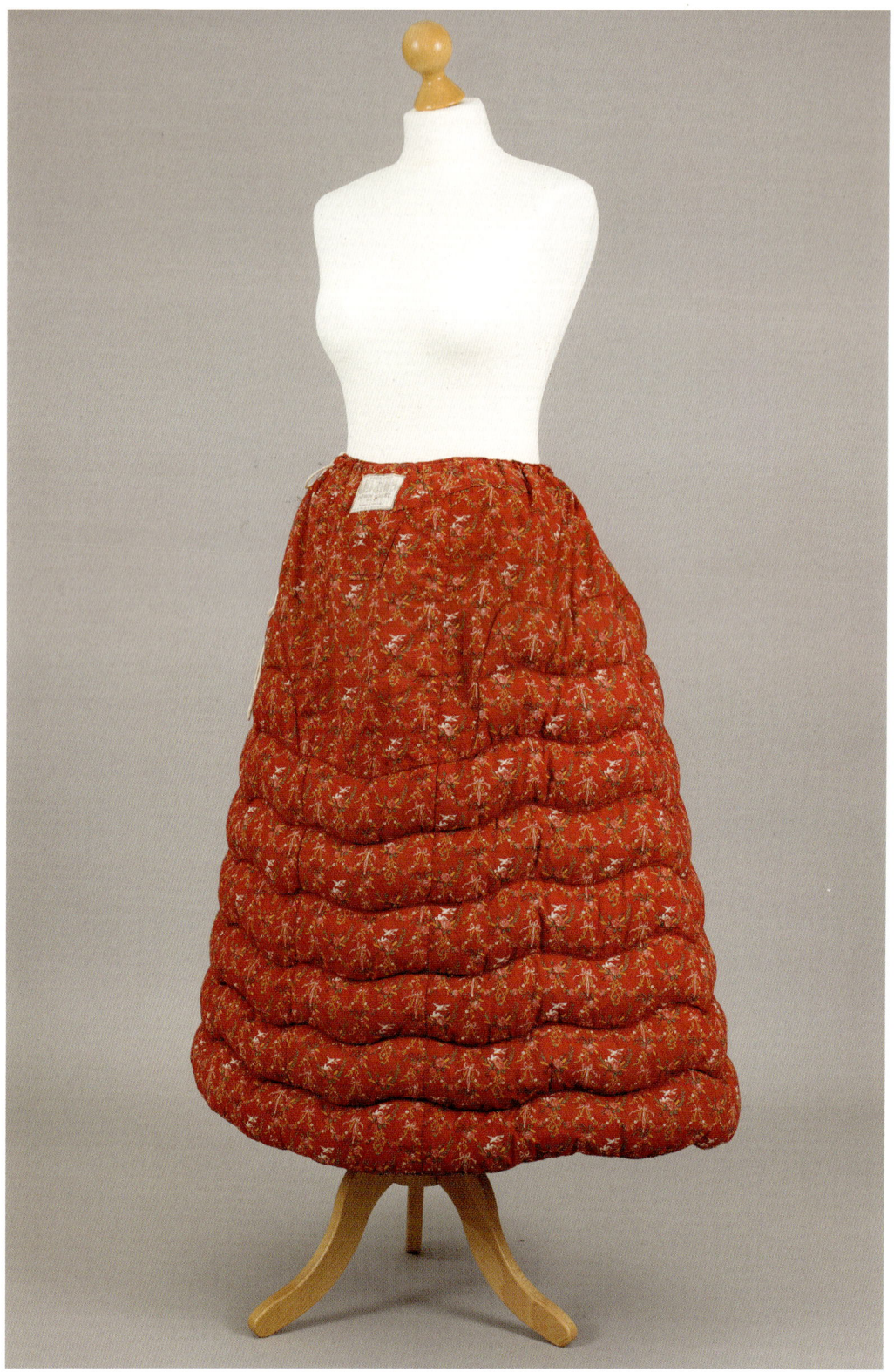

Figure 6.22 Booth and Fox quilted, filled petticoat from the late nineteenth century. *The Quilters' Guild Museum Collection*

Figure 6.23 An American child's dress made of printed Turkey red calico, *c.* 1864–5. *Cooper Hewitt, Smithsonian Design Museum. Dress (USA), 1864–65; printed cotton. 1954-116-3*

during the First World War shown in Figure 6.25, a woman wears a red apron and headscarf with a paisley pattern consistent with a Turkey red print.

Surviving objects and artworks indicate Turkey red was used in women's clothing in a variety of places over many decades. Relative to other printed fabrics it is not very common, probably due to the limited print palette and intense hue. Examples include a red cotton dress in Regency style, shown in Figure 6.26, made around 1815 in Barcelona and decorated with a black, red, and yellow printed scene at the bottom. Figure 6.27 shows a watercolor illustration of an 1820s dress from a North American collection in bright red with a fine floral pattern in the characteristic Turkey red

Figure 6.24 A child's coat made of printed Turkey red, *c.* 1850. American. *With permission of the Royal Ontario Museum © ROM. 975.241.105*

palette. A similar garment is found at the Victoria and Albert Museum (cat. no. T.74-1988), which has been analysed and confirmed as Turkey red. This dress has a concealed panel that unfastens at the waist, allowing the wearer to nurse an infant. In an 1860 watercolor in Figure 6.28, a Surinamese woman wears a turban and *koto misi* of bright red fabric with white polka dots of different sizes looking very much like Turkey red.

Lead-plate discharge-printed Turkey red with white and yellow patterns on red ground was used as garment material in the South Pacific in the late nineteenth and early twentieth centuries.[47] A swatch in Figure 6.29 shows such a pattern from the mid-nineteenth century depicting a white pineapple on the red ground flanked by a yellow border. In Figure 6.30, an 1891 painting (Ia Orana

Figure 6.25 Red cotton twill bathing costume, early 1900s.
© *Fashion Museum Bath / Bridgeman Images, dated 1904. I.21.34*
BATMC

Maria) by Paul Gaugin from when he lived in Tahiti depicts a woman in the foreground wearing a traditional garment called a *pareu* in bright red with a white pattern. It is likely this was made of imported Turkey red.

The collection of the American Museum of Natural History includes a red cotton women's dress with a fine white print (cat. no. H/15209) made around 1900 by the Laguna Pueblo tribe, and an Apache women's blouse with gathered sleeves and neck in a pattern of white dots (cat. no. H/15201). Also from around this time is a plain red cotton dress from Nisyros, Greece in the Victoria and Albert Museum collection (cat. no. T.149-1930). A colorized 1920s photograph of an Indian woman

Figure 6.26 Autochrome from 1916 by Jean-Baptiste Tournas-soud from the First World War of a French soldier (*poilu*) and a young woman. The apron and headscarf worn by the woman bear a pattern characteristic of printed Turkey red. *Commencement d'idylle, Oise, 1916. Jean-Baptiste Tournassoud/SPA/ ECPAD/Défense/AUL 110*

wearing a dancer's outfit in Figure 6.31 shows her large, full skirt in a similar paisley pattern on a red ground—possibly printed Turkey red from Scotland. Finally, a white bobbin lace cap from nineteenth-century Hungary, shown in Figure 6.32, appears to have been decorated across the bottom edge with a scrap of red cotton bearing a fine white print, adding a bit of brightness to an otherwise quotidian garment.

Figure 6.27 A red cotton dress from Barcelona *c.* 1815 with yellow and black printed accents. The fabric for this dress may have been made in France. *Website of the Museu del Disseny de Barcelona, museudeldisseny.barcelona.cat. MTIB 88027*

There is some evidence of Turkey red in men's wear, though it is not clear why there are comparatively fewer examples of red cotton men's clothing. Men were regular users of Turkey red bandanas, but whether other garments were uncommon or just have a low rate of preservation is unknown. One such item is an early twentieth-century red cotton men's shirt from the Yuchi tribe at the American Museum of Natural History (cat. no. 50/5824). The Henry Art Gallery has an early

Figure 6.28 A watercolor made in 1935 of an 1820s dress in bright red with a floral print in yellow, green, blue, and black characteristic of Turkey red. A similar example can be seen in the Victoria and Albert Museum collection (T.74-1988). *Courtesy National Gallery of Art, Washington. Virginia Berge, American, active C. 1935, Anonymous Craftsman (object maker), Plymouth Antiquarian Society (object owner), Dress, 1935/1942, Index of American Design. 1943.8.1574*

Figure 6.29 An 1860 watercolor of a Surinamese woman wearing a turban and *koto misi* while carrying a handkerchief, all with a distinct bright red and white pattern of Turkey red. *Rijksmuseum. Woman in koto misi from Album met tekeningen van Suriname. Jacob Marius Adriaan Martini van Geffen, 1860. RP-T-1994-281-35*

Figure 6.30 A sample of Turkey red printed for the South Pacific market showing a large white pineapple with a yellow border. These patterns continued to be printed with lead plates after production shifted to copper rollers. *University of Glasgow Archives and Special Collections, Records of United Turkey Red Co Ltd. GB248 UGD 13/8/5*

Figure 6.31 The French artist Paul Gaugin painted Ia Orana Maria (Hail Mary) in Tahiti in 1891. The red and white pareu worn by the woman in the foreground is characteristic of Turkey red made for the South Pacific market. *Metropolitan Museum of Art. Bequest of Sam A. Lewisohn, 1951. 51.112.2*

Figure 6.32 Colorized photograph of an Indian dancer, early 1920s, wearing a large pleated skirt in a print and colorway characteristic of Turkey red imported from Europe. *Peoples of All Nations. Charles Hose, 1922*

twentieth-century men's gym suit made of red cotton twill (cat. no. TC 82.1-171, t1 and t2), an unusual example of a garment likely to be made of Turkey red. A bale label design from the Archibald Orr Ewing company depicts a soldier holding a rifle and wearing a red tunic, white trousers, and a blue turban. A "Bombay Merchants' Record Book" from the same period describes a visit from agents who were buyers for military supplies in India who were "our largest buyers of Plain and Twill Turkey Red cloths." In a text on Indian army uniforms, a description for 1838 dress regulations for Madras engineers says "In warm weather . . . a blue cloak could be worn at walking length, lined with scarlet shalloon and clasp ornaments at the bottom of the collar." Shalloon is a term typically associated with a fine woolen fabric, but a late nineteenth-century Anglo-Indian dictionary says "Shalee, Shaloo, Shella, Sallo—names in familiar use for a soft twilled cotton stuff of a Turkey Red colour somewhat resembling what we call . . . shalloon."[48] No doubt a cotton lining would be more comfortable in the hot, humid climate. The University of Rhode Island Historic Textile and Costume Collection includes three mid-nineteenth century men's banyans, a stylish leisure garment for wearing around the home. Another association with menswear and Turkey red is the black and red prints sometimes referred to as "Garibaldi Cloth" in reference to Giuseppe Garibaldi, the Italian military commander and republican patriot who campaigned for Italian unification in the mid-nineteenth century. These were made by American manufacturers from about 1875 to 1925, but there is no consistent evidence to indicate that the red shirts worn by the *garibaldini* were Turkey red, as the movement was a voluntary one and had no official uniform.[49]

A few distinct examples of Turkey red use in objects include a set of Mahdiya horse armor from Sudan made in the late nineteenth century (see Figure 6.33), which comprises a quilted cotton cover of red, yellow, blue, and black triangles lined with fiber padding and toggle fasteners. The bright red cotton, along with possible lead chromate and Prussian blue, are characteristic of Turkey red prints. The Science Museum Group collection holds an Ashanti *akua mona* doll dated 1850–1920 (cat. no. A655908), which appears to be wearing a scrap of printed Turkey red cloth. These objects indicate Turkey red was commonly available across Africa by the late nineteenth century. Finally, although faded where the photographer's head was positioned, a red cotton photographer's hood from the turn of the twentieth century in the Powerhouse Museum collection (cat. no. 97/48/1) remains bright enough to imply it may be made of Turkey red.

Conclusion

The documentary and object record, particularly for the second half of the nineteenth century, presents a fascinating and varied picture of Turkey red use. It must be acknowledged that the available information does not necessarily present a complete history of Turkey red, and that very little survives from before the nineteenth century. Publications, particularly periodicals, document trade in Turkey red and the newly-formed industries. Although it was considered to be of a higher quality, Turkey red was affordable for working people, particularly after the mechanization of textile production significantly lowered prices and increased access to the market.

Figure 6.33 A bobbin lace cap trimmed with Turkey red, nineteenth century. Hungary. *Clevelad Museum of Art. Gift from J. H. Wade. 1923.459*

Figure 6.34 Horse armor made from quilted cotton and fiber padding. Mahdiya culture, Sudan, late nineteenth century. © *Trustees of the British Museum. Af1899,1213.2*

Using the definition of Turkey red given at the beginning of this book, a number of objects in collections have been tentatively identified as Turkey red. While it is important to have robust confirmation in some cases, in reality this is often limited. Red cotton in objects made before the 1920s, particularly those which have been printed with the distinctive palette or which have seen heavier use, can be reasonably considered to be Turkey red. The objects discussed here include domestic towels and bedding, dresses and accessories for women, men's shirts and uniforms, bandanas for all ages, and useful or decorative items. Some are likely Turkey red for its particular hue, for its durability, or for both. Although it represents a small portion of textiles in collections overall, Turkey red is consistently present in many items made over decades probably because of these particular features.

Conclusions

Turkey red is a fascinating case study of the intersections between art and science, craft and industry, and trade and knowledge exchange. Until the early twentieth century, it was generally considered to be not only the *best*, but the *only* way to obtain a good, fast red on cotton or linen fibers. Turkey red was superlative, particularly renowned for its brilliant hue and fastness to washing, light, rubbing, and bleaching. It was also notoriously difficult to make, but its history shows it was valued enough to still be a worthwhile endeavor. Prior to the 1870s, Turkey red was dyed in a lengthy and complicated process that took around a month to complete. The value of the industry drove technological innovations in material research and development to improve production, resulting in Turkey red oil, the new process, and more efficient use of time and materials. It also motivated the development of madder concentrates and then synthetic alizarin, which became the first naturally-occurring dye replicated in a laboratory and replaced madder in Turkey red. This was roughly concurrent with the transition to the new process. Although these developments revolutionized Turkey red production, they fulfilled the same material function as the original materials.

The fundamental chemistry of Turkey red dyeing begins with a fatty acid layer deposited onto cellulose fibers via hydrogen bonding. This was accomplished in the old process using a bath made of rancid plant oils and an alkaline ley; the new process used manufactured Turkey red oil and water. Not only was the oil treatment essential for Turkey red, its position as the initial step appears to be unique to the process. The old process often included ruminant dung in the bath, which assisted the process but was not part of the final complex. After oiling, the fibers were treated with aluminium, which complexed with the fatty acids to deposit aluminium soaps onto the fibers. Their low water solubility contributes to the wash fastness of the final color, as do the hydrophobic chains of the fatty acids which create some water resistance. A tannin treatment may also be applied before the aluminium bath; this was also said to improve the result but was not essential to the process. The Turkey red complex forms in the dye bath when anthraquinone dyes attach to the aluminium ions with calcium ions between them. During this step, the white fabric becomes patchy pink as the dye bath darkens, eventually emerging a deep crimson color. After clearing in a boiling soap or oil treatment, the excess color is removed and the Turkey red is finished.

The origins of Turkey red may never be confirmed, but available evidence indicates it is Indian. Historical trade records and documents show red cotton was traded from India since at least the thirteenth century, and the subcontinent has a rich textile history, further supporting this conclusion. Unfortunately, there is little information on the extent of Indian Turkey red in collections and no objects were identified during this research. The definition of Turkey red used here makes it possible to trace similar practice in Indonesia and the Eastern Mediterranean, where knowledge of the process spread. From the eighteenth century, European textile manufacturers, professional societies, and governments made a consistent effort to obtain practical knowledge of Turkey red dyeing and establish a domestic industry, eliminating the need to import. This effort was concurrent with the development of new machinery in the Industrial Revolution and the mechanization of textile production. The exponential increase in production capacity demanded more raw material than ever before, and a growing market for cotton textiles contributed to the trade in enslaved people and depended heavily on their labor. Learned minds sought to understand the chemistry of Turkey red dyeing and how it could be improved. Research on the materials and process was facilitated by significant advances in the field of chemistry during the nineteenth century, and European manufacturing made it possible to export Turkey red on a global scale. Major industries grew in Scotland, England, France, and Switzerland, and undermined hand production in India and the Eastern Mediterranean, which became the export markets for their former customers.

The object record for Turkey red is probably not representative of how it was made, used, and disseminated due to a number of factors. In general, textiles are particularly susceptible to degradation by light and moisture. Items that end up in collections have a higher survival rate, but collections are usually biased according to individual or institutional taste. Turkey red was a premium product in the sense that it cost more than plain cotton, but it was not so expensive that it was a luxury out of reach for most. As such, it is not often found in many higher-status objects that are more often found in collections. Because of its durability, particularly against washing and bleaching, it was valued as a decorative element in quotidian objects like towels, blankets, tablecloths, and serviettes. It was also used in more heavily used garments, like children's' clothing and swimwear, and appears in adult garments and accessories of various provenances.

This book explores the history, chemistry, and use of Turkey red textiles through a process- and material-based definition. Many of the conclusions depend on the assumption that prior to the early twentieth century, red cotton was likely to be Turkey red. The scope of this work is greater than other studies on the topic, but far from comprehensive. There may be much more to learn about the industries in France and Switzerland, about dyeing outside of Europe, and about how Turkey red was traded and used.

Glossary

acid More hydrogen ions (H+) than hydroxide (-OH) are present. See pH.

alkaline hydroxide (OH–) ions are present. See pH.

aniline molecule in coal tar that formed the basis of many early synthetic dyes

aqueous a solution for which the solvent is water

bleaching powder see calcium hypochlorite (in Chemicals list)

carbonyl carbon double bonded to oxygen, C=O

carboxylic acid -COOH functional group

cellulose the chemical basis of cotton and linen fibers, composed of linked d-glucose rings

chalk see calcium carbonate (in Chemicals list)

chrome green see chromium (III) oxide (in Chemicals list)

Congo Red early synthetic red dye that resembled Turkey red visually, but had very poor fastness

covalent bond formed when two atoms share an electron pair and are very stable. These bonds are present between atoms in a material like cotton. Covalent bonds exist in water-soluble compounds like table sugar (sucrose), however its solubility is based on the small size of sucrose relative to cellulose and not the disruption of the covalent bonds within the molecule

crocking color transfer by rubbing

cyano group carbon triple bonded to nitrogen, C≡N

dye organic molecule with chromogen that generates visible color based on alternating single and double bonds, and functional groups like hydroxyls and carbonyls

dyeing the application of dyes to textile fibers, often from an aqueous solution. A comprehensive examination of dyeing in the modern era through an economic and technological perspective can be found in *Engel's Farben der Globalisierung: die Entstehung moderner Märkte für Farbstoffe 1500–1900* and Brunello's *The Art of Dyeing in the History of Mankind*.

dyeworks a mid- to large-sized textile dyeing operation

fastness the stability of a color to a variety of conditions including light exposure, washing, rubbing, and bleaching

flocculation the appearance of particles in solution that do not precipitate

free fatty acid a fatty acid no longer bound to glycerol

functional group substituent of a molecule that causes characteristic chemical reactions, like hydroxyl, carboxylic acid, etc.

fustian linen-cotton weave

garancine mid-nineteenth century madder product made by treating ground root with sulfuric acid to reduce impurities and increase dyeing power

hydrocarbon a molecule consisting entirely of carbon and hydrogen

hydrogen bond an electrostatic attraction, form between hydrogen and oxygen, nitrogen, and fluorine atoms. Their strength is generally less than ionic and covalent bonds.

hydrolyse to break a bond with water

hydrophobic resists or does not mix well with water

Industrial Revolution period from roughly 1760–1840 during which textile manufacturing transitioned from hand to machine production.

inorganic materials that are not carbon-based, though they may still contain carbon, like rocks, minerals, and metals.

ionic bond formed through electrostatic attraction, between positive and negative charges. A familiar example is table salt, sodium chloride (NaCl), where each has a net charge of one (Na+ and Cl–) and the opposites attract. These bonds are often broken in the presence of water, which is why salt will dissolve but cotton does not.

Ions atoms or molecules carrying a positive or negative charge

isomer a molecule with the same formula but a different structure, e.g. purpurin and isopurpurin

kalamkari traditional Indian textile printing technique with material similarity to Turkey red

Levant, The an Eastern Mediterranean region with a variable definition that loosely includes Egypt, Lebanon, Jordan, Israel, Palestine, Syria, and a large part of Turkey.

level even color on dyed fibers

ley an aqueous alkaline solution often made with plant ash

madder plant from the genus Rubiaceae whose root was used in Turkey red dyeing, primarily *Rubia tinctorum* L.

mauveine the first fully-synthetic dye derived from coal tar, made by William Henry Perkin in 1856.

molecules larger units of matter with varying degrees of complexity and comprised of atoms

mordant a metal treatment on fibers to facilitate the bonding of dye molecules, particularly for natural dyes

oleum mixture of sulfuric acid and sulfur trioxide used in alizarin synthesis and Turkey red oil manufacture

organic materials with a carbon structure and carbon-hydrogen bonds. Fibers like wool and cotton are organic materials, as manufactured ones like polyester

pH The pH scale, which measures hydrogen ion concentration, was developed in the early twentieth century. A neutral value on the pH scale is 7. Increasingly alkaline materials have a higher pH going up to 14, and increasingly acidic materials have a lower pH. Precise pH measurements are not a part of dyeing Turkey red, but the distinction between acidic and alkaline conditions is relevant. Under acidic conditions, the concentration of hydrogen ions (H+) is greater, while under alkaline conditions the concentration of hydroxide (OH–) ions is greater.

piece dyeing dyeing whole cloth

pigment physical particles of colored material, typically held in a binding medium like oil or gum.

polychrome decorated with multiple colors

polysaccharide long polymeric compounds composed of monosaccharide, or simple sugar units

precipitate a solid that forms when two solutions react to yield an insoluble product that settles out as particles

printworks a mid- to large-sized textile printing operation

Prussian blue a dark blue pigment produced by oxidation of ferrous ferrocyanide salts. See ferric ferrocyanide (in Chemicals list)

pullicate checked woven cloth

quicklime made by heating limestone (a rock rich in calcium carbonate) in a kiln, commonly used as a source of alkali. See calcium oxide (in Chemicals list)

scouring before a fiber can be dyed or printed, the natural oils, fats, waxes, dirt, and other materials on the fiber that would interfere by preventing the colorants from attaching to the textile must be removed. For cotton, the raw fibers are coated with wax, pectin, and oil. Boiling it in an aqueous alkaline bath removes this extraneous material, cleansing the fibers. The scoured cotton whitens and the bath becomes dark with the dissolved matter. Scouring cannot be replaced by washing with soap since it is not caustic enough to effectively remove everything. Fibers may also be bleached before dyeing to achieve a white ground.

soda see sodium carbonate (in Chemicals list)

tin crystals see tin (II) chloride (in Chemicals list)

tinctorial the ability to dye

Turkey red oil the first synthetic anionic surfactant, developed specifically for the Turkey red industry in the 1870s and made by treating castor oil with sulfuric acid, hence its occasional name "sulfated castor oil"

unsaturated a hydrocarbon containing at least one double bond

Chemicals

alizarin: 1,2-dihydroxy anthraquinone
alum: formally potassium alum, $KAl(SO_4)_2$
aluminium hydroxide: $Al(OH)_3$
aluminium phosphate: $AlPO_4$
ammonium: NH_4^+
anthrapurpurin: 1,2,7-trihydroxy anthraquinone
anthraquinone: 9,10-dioxoanthracene
calcium carbonate: $CaCO_3$
calcium hydroxide: $Ca(OH)_2$
calcium hypochlorite: $Ca(ClO)_2$
calcium oxide: CaO
chromium (III) oxide: Cr_2O_3
ferric ferrocyanide: Prussian blue, $Fe(III)_4[Fe(II)(CN)_6]_3$
flavopurpurin: 1,2,6-trihydroxy anthraquinone
gypsum: calcium sulfate, $CaSO_4$
hydrochloric acid: HCl
hydroxyl: -OH functional group
iron sulfate: $FeSO_4$
lead acetate: sugar of lead, $Pb(CH_3COO)_2$
lead chromate: $PbCrO_4$
lime: CaO and/or $Ca(OH)_2$
naphthalene: two-ring $C_{10}H_8$
potassium chlorate: $KClO_3$
potassium dichromate: K_2Cr2O_7
potassium hydroxide: KOH
Prussian blue: see ferric ferrocyanide
purpurin: 1,2,4-trihydroxy anthraquinone
sodium arsenate: Na_3AsO_4
sodium carbonate: Na_2CO_3
sodium hydroxide: $NaOH$
sodium hypochlorite: $NaClO$
sodium phosphate: Na_3PO_4
sulfate: $-OSO_3$

sulfonic acid: -SO_3^- functional group
sulfur trioxide: SO_3
sulfuric acid: H_2SO_4
tin (II) chloride: $SnCl_2$
sodium sulfate: Glauber's salt, Na_2SO_4

Notes

Preface

1. Jacqué et al., *Andrinople: Le rouge magnifique*.
2. Nenadic and Tuckett, *Colouring the Nation: The Turkey Red Printed Cotton Industry in Scotland c.1840-1940*.

Introduction

1. Ure, *A Dictionary of Arts, Manufactures, and Mines*; Carruthers, "A New Process of Dyeing Turkey Red."
2. Jacqué et al., *Andrinople: Le rouge magnifique*.
3. Andriessen, "Turkish Red & More"; Boot, "Turkish Red & More : Formafantasma."
4. Nenadic and Tuckett, *Colouring the Nation: The Turkey Red Printed Cotton Industry in Scotland c. 1840–1940*.
5. Brackman, "Turkey Red Tour"; "First Friday + Barbara Brackman's Lecture—The Quilt Detective: Clues in Turkey Red"; Bamford, "Turkey Red."
6. Wertz, "Turkey Red Dyeing in Late-19th Century Glasgow: Interpreting the Historical Process through Re-Creation and Chemical Analysis for Heritage Research and Conservation."
7. The National Archives, "The National Archives Currency Converter"; Official Data Foundation, "Official Data Foundation."

Chapter 1

1. Sansone, "Alizarin-Red and Turkey-Red Dyeing and Printing on Cotton."
2. Knecht, Rawson, and Loewenthal, *A Manual of Dyeing*, 2:588–603.
3. Brackman, *Clues in the Calico: A Guide to Identifying and Dating Antique Quilts*; Liles, *The Art and Craft of Natural Dyeing: Traditional Recipes for Modern Use*.
4. Verbong, "De ontwikkeling van het stakingsrecht in Nederland."
5. Peel, "Turkey Red Dyeing in Scotland Its Heyday and Decline."
6. "Sunday School Object Lesson."
7. Arthur et al., *Seeing Red: Scotland's Exotic Textile Heritage*, 19.

8. Rauch, *John Rauch's Receipts on Dyeing, in a Series of Letters to a Friend. Containing Correct and Exact Copies of All His Best Receipts on Dyeing Cotton and Woollen Goods; Obtained and Improved by Him, during Twelve Years Practice, at Different Manufactories in Switzerland, France, Germany and America. Also, a true Description of His Invented Substitute for Woad, Being a Cheap and Preferable Material, and the Produce of this Country*, 34; Dumas, *Traité de chimie, appliquée aux arts*; Ure, *A Dictionary of Arts, Manufactures, and Mines*; Sansone, "Alizarin-Red and Turkey-Red Dyeing and Printing on Cotton."

9. Arthur et al., *Seeing Red: Scotland's Exotic Textile Heritage*, 22.

10. Dépierre, *Traité de la teinture et de l'impression des matières colorantes artificielles. 2me partie : L'Alizarine artificielle et ses dérivés*, 2:243–4; Le Pileur d'Apligny, *L'art de la teinture des fils et étoffes de coton*; Vitalis, *Cours élémentaire de teinture sur laine, soie, lin, chanvre et coton, et sur l'art d'imprimer les toiles*; Persoz, *Traité théorique et pratique de l'impression des tissus*.

11. Kendall, "Preparation of Alizarine Assistant and Its Action in Turkey-Red Dyeing."

12. Schaefer, "The Cultivation of Madder."

13. Sansone, "Alizarin-Red and Turkey-Red Dyeing and Printing on Cotton"; Thomson, "I. New Theory of Dyeing Turkey Red"; Washington, "On Dyeing Turkey Red."

14. Berthollet, *Elements of the Art of Dyeing*, 144; O'Connor, "Four Aspects of Turkey Red: The Turkey Red Industry: Export Cloths."

15. Kiel and Heertjes, "Metal Complexes of Alizarin V. Investigations of Alizarin–Dyed Cotton Fabrics."

16. Tarrant, "The Turkey Red Dyeing Industry in the Vale of Leven."

17. Schaefer, "The Cultivation of Madder."

18. Clibbens, "Some Research Problems in Cotton Bleaching and Dyeing," 249.

19. Hofenk de Graaff, Roelofs, and van Bommel, *The Colourful Past*, 96.

20. Burns, "Parish of Barony of Glasgow"; Ure, *A Dictionary of Arts, Manufactures, and Mines*.

21. Hellot, Macquer, and Le Pileur d'Apligny, *The Art of Dyeing Wool, Silk, and Cotton*, 467.

22. Wilson, *As Essay on Light and Colours, and What Colouring Matters Are That Dye Cotton and Linen*.

23. Washington, "On Dyeing Turkey Red."

24. Liles, *The Art and Craft of Natural Dyeing: Traditional Recipes for Modern Use*; Fieser, "The Discovery of Synthetic Alizarin."

25. Tarrant, "The Turkey Red Dyeing Industry in the Vale of Leven."

26. Pallas, "I. The Genuine Oriental Process for Giving to Cotton Yarn or Stuff the Fast or Ingrained Colour, Known by the Name of Turkey Red, as Practised at Astracan."

27. MacFarlan et al., "City of Glasgow and Suburban Parishes of Barony and Gorbals."

28. Sansone, "Alizarin-Red and Turkey-Red Dyeing and Printing on Cotton."

29. Wilson, *As Essay on Light and Colours, and What Colouring Matters Are That Dye Cotton and Linen*.

30. Sansone, "Alizarin-Red and Turkey-Red Dyeing and Printing on Cotton."

31. Rauch, *John Rauch's Receipts on Dyeing, in a Series of Letters to a Friend. Containing Correct and Exact Copies of All His Best Receipts on Dyeing Cotton and Woollen Goods; Obtained and Improved by Him, during Twelve Years Practice, at Different Manufactories in Switzerland, France, Germany and America. Also, a true Description of His Invented Substitute for Woad, Being a Cheap and Preferable Material, and the Produce of this Country*, 34; "Dyeing Turkey Red"; Sansone, "Alizarin-Red and Turkey-Red Dyeing and Printing on Cotton"; Schaefer, "The History of Turkey Red Dyeing"; Wilson, *As Essay on Light and Colours, and What Colouring Matters Are That Dye Cotton and Linen*; Nenadic, "Selling Printed Cottons in Mid-Nineteenth-Century India: John Matheson of Glasgow and Scottish Turkey Red."

32. Verbong, "Turksrood," 271–2.

33. D.B., "On Dyeing Red"; Sansone, "Alizarin-Red and Turkey-Red Dyeing and Printing on Cotton."

34. O'Neill and Fesquet, *A Dictionary of Dyeing and Calico Printing*, 338.

35. Nieto-Galan, *Colouring Textiles : A History of Natural Dyestuffs in Industrial Europe.*

36. Cain and Thorpe, *The Synthetic Dyestuffs and the Intermediate Products from Which They Are Derived*; Carruthers, "A New Process of Dyeing Turkey Red."

37. Trotman, *Dyeing and Chemical Technology of Textile Fibres*, 458.

38. O'Neill, "The Printing and Dyeing of Calico, Silk, and Woollen Fabrics," 358.

39. Fairlie, "Dyestuffs in the Eighteenth Century"; O'Neill and Fesquet, *A Dictionary of Dyeing and Calico Printing*, 159–61.

40. Clow and Clow, "The Chemical Revolution: A Contribution to Social Technology," 271; O'Neill and Fesquet, *A Dictionary of Dyeing and Calico Printing.*

41. "Correspondence with Marta Turok."

42. Potukuchi, "The World of the Weaver in the Northern Coromandel, 1750–1850."

43. Brackman, *Clues in the Calico: A Guide to Identifying and Dating Antique Quilts*, 62–3.

44. Smith, "Four Aspects of Turkey Red."

45. Dumas, *Traité de chimie, appliquée aux arts*, 8:411; Chateau, "Critical and Historical Notes Concerning the Production of Adrianople or Turkey Red, and the Theory of This Colour No. 10," 139.

46. Faroqhi, "Ottoman Cotton Textiles, 1500s to 1800: The Story of a Success That Did Not Last."

Chapter 2

1. Vitalis, *Cours élémentaire de teinture sur laine, soie, lin, chanvre et coton, et sur l'art d'imprimer les toiles*; Persoz, *Traité théorique et pratique de l'impression des tissus.*

2. Cunningham et al., "Hanging by a Thread: Natural Metallic Mordant Processes in Traditional Indonesian Textiles."

3. Koren, "The Colors and Dyes on Ancient Textiles in Israel."

4. Dumas, *Traité de chimie, appliquée aux arts.*

5. Neyland, "The Seagoing Vessels on Dilmun Seals."

6. Beaujard, "Gujarat and Long-Distance Trade in the Indian Ocean Region before the Sixteenth Century."

7. Scott et al., "Exotic Foods Reveal Contact between South Asia and the Near East during the Second Millennium BCE."

8. Macquer, *Dictionnaire portatif des arts et métiers: contenant en abrégé l'histoire, la description & la police des arts et métiers, des fabriques et manufactures de France et des pays étrangers*, 3:485.

9. Reinking and Atayolu, "Zur Entstehung und Frühgeschichte des Türkischrots."

10. Ashmore et al., *The Fabric of India*; Fotheringham, *The Indian Textile Sourcebook.*

11. Moulherat et al., "First Evidence of Cotton at Neolithic Mehrgarh, Pakistan: Analysis of Mineralized Fibres from a Copper Bead."

12. Gillard et al., "The Mineralization of Fibres in Burial Environments."

13. Chao, *Chau Ju-Kua: His Work on the Chinese and Arab Trade in the Twelfth and Thirteenth Centuries, Entitled Chu-Fan-Chï.*

14. The Editors of Encyclopaedia Britannica, "Zhao Rukuo."
15. Chao, *Chau Ju-Kua: His Work on the Chinese and Arab Trade in the Twelfth and Thirteenth Centuries, Entitled Chu-Fan-Chï*, 126.
16. Beaujard, "Gujarat and Long-Distance Trade in the Indian Ocean Region before the Sixteenth Century."
17. Mohanty, Chandramouli, and Naik, *Natural Dyeing Processes of India*, 18.
18. Mohanty, Chandramouli, and Naik, 271.
19. Schmitt et al., "Aluminium Accumulation and Intra-Tree Distribution Patterns in Three Arbor Aluminosa (Symplocos) Species from Central Sulawesi."
20. Driessen, "Étude sur le rouge turc, ancien procédé"; Jansen, Watanabe, and Smets, "Aluminium Accumulation in Leaves of 127 Species in Melastomataceae, with Comments on the Order Myrtales"; de Niederhäusern, "Rapport sur le travail de M. F. Driessen: 'Etude sur le rouge turc, ancien procédé', et sur le contenu de deux plis cachetés Nos. 700 et 1276, déposés par le même auteur."
21. Mukund, "Indian Textile Industry in 17th and 18th Centuries: Structure, Organisation and Responses."
22. Guy, "One Thing Leads to Another: Indian Textiles and the Early Globalization of Style."
23. Knecht, Rawson, and Loewenthal, *A Manual of Dyeing*, 1893, 1:174; Baker, "East Indian Hand-Painted Calicoes of the Seventeenth and Eighteenth Centuries, and Their Influence on the Tinctorial Arts of Europe."
24. Mukund, "Indian Textile Industry in 17th and 18th Centuries: Structure, Organisation and Responses."
25. Heyne, *Tracts, Historical and Statistical, on India: With Journals of Several Tours Through Various Parts of the Peninsula : Also, an Account of Sumatra, in a Series of Letters*, 204–18.
26. Mohanty, Chandramouli, and Naik, *Natural Dyeing Processes of India*, 11–12.
27. Robinson, *A History of Dyed Textiles*, 81.
28. Hanna, *Ottoman Egypt and the Emergence of the Modern World: 1500–1800*.
29. Baker, "East Indian Hand-Painted Calicoes of the Seventeenth and Eighteenth Centuries, and Their Influence on the Tinctorial Arts of Europe," 482–83.
30. Beaujard, "Gujarat and Long-Distance Trade in the Indian Ocean Region before the Sixteenth Century."
31. Rouffaer and Juynboll, *De batik-kunst in Nederlandsch-Indië en haar geschiedenis*, 334; Verbong, "Technische innovaties in de katoendrukkerij en -ververij in Nederland 1835–1920," 188; Cunningham et al., "Hanging by a Thread: Natural Metallic Mordant Processes in Traditional Indonesian Textiles."
32. Crookes, *A Practical Handbook of Dyeing and Calico-Printing*, 327–8.
33. Cunningham et al., "Hanging by a Thread: Natural Metallic Mordant Processes in Traditional Indonesian Textiles."
34. Faroqhi, "Ottoman Cotton Textiles, 1500s to 1800: The Story of a Success That Did Not Last"; Dumas, *Traité de chimie, appliquée aux arts*; Chateau, *Étude historique et chimique pour servir à l'histoire de la fabrication du rouge turc ou d'Andrinople et à la théorie de cette teinture*; Gekas, "A Global History of Ottoman Cotton Textiles, 1600–1850"; Adams, "Clothing and Textiles of Ottoman Egypt: Examples from Art and Archaeology"; Atayolu, "Beitrage zur Geschichte der Farberei aus turkischen Archiven"; Raveux, "The Orient and the Dawn of Western Industrialization: Armenian Calico Printers from Constantinople in Marseilles (1669-1686)";

Fukasawa, *Toile et commerce du Levant d'Alep à Marseille*; Girard, *Mémoire sur l'agriculture, l'industrie et le commerce de l'Égypte*; Hanna, *Ottoman Egypt and the Emergence of the Modern World: 1500–1800*; Keyder, Özveren, and Quataert, "Port-Cities in the Ottoman Empire: Some Theoretical and Historical Perspectives"; Reinking and Atayolu, "Zur Entstehung und Frühgeschichte des Türkischrots"; Faroqhi, "Declines and Revivals in Textile Production"; MacDonald, "Spanish Textile and Clothing Nomenclature in -án, -í, and -ín"; Pallas, "I. The Genuine Oriental Process for Giving to Cotton Yarn or Stuff the Fast or Ingrained Colour, Known by the Name of Turkey Red, as Practised at Astracan"; Verbong, "Technische innovaties in de katoendrukkerij en -ververij in Nederland 1835–1920"; Katsiardi-Hering, "The Allure of Red Cotton Yarn, and How It Came to Vienna: Associations of Greek Artisans and Merchants Operating between the Ottoman and Hapsburg Empires."

35. Reinking and Atayolu, "Zur Entstehung und Frühgeschichte des Türkischrots."

36. Katsiardi-Hering, "The Allure of Red Cotton Yarn, and How It Came to Vienna: Associations of Greek Artisans and Merchants Operating between the Ottoman and Hapsburg Empires"; Gekas, "A Global History of Ottoman Cotton Textiles, 1600-1850."

37. Faroqhi, "Ottoman Cotton Textiles, 1500s to 1800: The Story of a Success That Did Not Last."

38. Masson, *Histoire du commerce français dans le Levant au XVIIIe siècle.*

39. Vitalis, *Cours élémentaire de teinture sur laine, soie, lin, chanvre et coton, et sur l'art d'imprimer les toiles*, 235.

40. Masson, *Histoire du commerce français dans le Levant au XVIIIe siècle*; Schaefer, "The History of Turkey Red Dyeing"; Hanna, *Ottoman Egypt and the Emergence of the Modern World: 1500–1800*; Miquelon, *Dugard of Rouen: French Trade to Canada and the West Indies, 1729–1770*, 129–30; Riello, *Cotton: The Fabric That Made the Modern World*, 179; Tarrant, "The Turkey Red Dyeing Industry in the Vale of Leven"; Cliffe, "Turkey Red in Blackley: A Chapter in the History of Dyeing"; "Mémoire contenant le procédé de la teinture du coton rouge-incarnat d'Andrinople sur le coton filé"; Flachat, *Observations sur le commerce et sur les arts d'une partie de l'Europe, de l'Asie, de l'Afrique, et même des Indes Orientales*, 2:405–14; Cardon, *Natural Dyes: Sources, Tradition, Technology and Science*; Conseil d'Etat France, "Arrêt du conseil d'état qui, en autorisant la manufacture de draps de soie, laine, ratines et peluches, établie à Montmartre par le Sieur Quinquet, lui permet de faire teindre dans ladite manufacture, en grand et bon teint et en rouge d'Andrinople, toutes."

41. Lopez, "The Transition from Natural Madder to Synthetic Alizarine in the American Textile Industry, 1870–1890."

42. Bremner, *The Industries of Scotland: Their Rise, Progress, and Present Condition.*

43. Robinson, *A History of Printed Textiles.*

44. Stirling, "History of Colour Printing in the United Kingdom."

45. "The Red Society of Glasgow: Minute Book."

46. "The Red Society of Glasgow: Seal of Cause from the Magistrates and Town Council of Glasgow to James Bogle, Preces, David Fairie, Collector, James Campbell, Alexander Scott, Gershom Robertson, James Marshall, William Killoch, William McFarland, William Fin."

47. Long, "Introduction"; Clow and Clow, "The Chemical Revolution: A Contribution to Social Technology."

48. Wadsworth and De Lacy Mann, *The Cotton Trade and Industrial Lancashire, 1600–1780*, 179–80.

49. Wilson, *As Essay on Light and Colours, and What Colouring Matters Are That Dye Cotton and Linen.*

50. Musson and Robinson, "Science and Technology in the Industrial Revolution," 344; Cliffe, "Turkey Red in Blackley: A Chapter in the History of Dyeing."
51. Tarrant, "The Turkey Red Dyeing Industry in the Vale of Leven."
52. Clow and Clow, "The Chemical Revolution: A Contribution to Social Technology," 215–16.
53. Beke, *Notes and Queries: A Medium of Intercommunication for Literary Men, General Readers, Etc.*; Huguenot Society of London, *Proceedings of the Huguenot Society of London*, 1:298–9.
54. Hofenk de Graaff, Roelofs, and van Bommel, *The Colourful Past*, 94.
55. O'Neill, "The Printing and Dyeing of Calico, Silk, and Woollen Fabrics."
56. Crookes, *A Practical Handbook of Dyeing and Calico-Printing*, 228.
57. Cardon, *Natural Dyes: Sources, Tradition, Technology and Science*, 119; Hofenk de Graaff, Roelofs, and van Bommel, *The Colourful Past*, 94.
58. Chenciner, *Madder Red: A History of Luxury and Trade*.
59. Blackburn, "Natural Dyes in Madder (Rubia Spp.) and Their Extraction and Analysis in Historical Textiles."
60. Schaefer, "The Cultivation of Madder."
61. Crookes, *A Practical Handbook of Dyeing and Calico-Printing*; Fairlie, "Dyestuffs in the Eighteenth Century"; Travis, "Between Broken Root and Artificial Alizarin: Textile Arts and Manufactures of Madder"; Schaefer, "The Cultivation of Madder."
62. Travis, "Between Broken Root and Artificial Alizarin: Textile Arts and Manufactures of Madder"; Cuoco et al., "Characterization of Madder and Garancine in Historic French Red Materials by Liquid Chromatography-Photodiode Array Detection"; Schaefer, "The Cultivation of Madder."
63. Miller, *The Method of Cultivating Madder, as It Is Now Practised by the Dutch in Zealand*.
64. Schaefer, "The Cultivation of Madder."
65. Lopez, "The Transition from Natural Madder to Synthetic Alizarine in the American Textile Industry, 1870–1890."
66. Travis, "Between Broken Root and Artificial Alizarin: Textile Arts and Manufactures of Madder."
67. Ure, *A Dictionary of Arts, Manufactures, and Mines*.
68. Schaefer, "The Cultivation of Madder."
69. Ure, *A Dictionary of Arts, Manufactures, and Mines*; Schaefer, "The Cultivation of Madder"; Travis, "Between Broken Root and Artificial Alizarin: Textile Arts and Manufactures of Madder."
70. Blackburn, "Natural Dyes in Madder (Rubia Spp.) and Their Extraction and Analysis in Historical Textiles."
71. Burnett and Thomson, "Naturally Occurring Quinones. Part XV. Biogenesis of the Anthraquinones in Rubia Tinctorum L. (Madder)"; Fieser, "The Discovery of Synthetic Alizarin"; Hill and Richter, "Anthraquinone Colouring Matters: Galiosin; Rubiadin Primeveroside."
72. Dépierre, *Traité de la teinture et de l'impression des matières colorantes artificielles. 2me partie : L'Alizarine artificielle et ses dérivés*, 2:40.
73. Schweppe and Winter, "Madder and Alizarin," 112.
74. Fieser, "The Discovery of Synthetic Alizarin."
75. Travis, "Between Broken Root and Artificial Alizarin: Textile Arts and Manufactures of Madder"; Travis, *The Rainbow Makers: The Origins of the Synthetic Dyestuffs Industry in Western Europe*.
76. Crookes, *A Practical Handbook of Dyeing and Calico-Printing*, 233.
77. Ure, *A Dictionary of Arts, Manufactures, and Mines*.
78. Decaisne, *Recherches anatomiques et physiologiques sur la garance*; Decaisne, "XXIX.—On the Root of the Madder."

79. Hofenk de Graaff, Roelofs, and van Bommel, *The Colourful Past*, 97–98.

80. Hobson and Wales, "'Green' Dyes."

81. Cooksey and Dronsfield, "Edward Schunck: Forgotten Dyestuffs Chemist?"

82. Cuoco et al., "Characterization of Madder and Garancine in Historic French Red Materials by Liquid Chromatography-Photodiode Array Detection."

83. Derksen et al., "Two Validated HPLC Methods for the Quantification of Alizarin and Other Anthraquinones in Rubia Tinctorum Cultivars."

84. Blackburn, "Natural Dyes in Madder (Rubia Spp.) and Their Extraction and Analysis in Historical Textiles"; Derksen et al., "Chemical and Enzymatic Hydrolysis of Anthraquinone Glycosides from Madder Roots"; Schaefer, "The Cultivation of Madder"; Travis, "Between Broken Root and Artificial Alizarin: Textile Arts and Manufactures of Madder."

85. Ure, *A Dictionary of Arts, Manufactures, and Mines.*

86. Colin and Robiquet, "Nouvelles Recherches sur la matière colorante de la garance"; Travis, *The Rainbow Makers: The Origins of the Synthetic Dyestuffs Industry in Western Europe.*

87. Travis, "Between Broken Root and Artificial Alizarin: Textile Arts and Manufactures of Madder"; Crookes, *A Practical Handbook of Dyeing and Calico-Printing*, 258–59; Knecht, Rawson, and Loewenthal, *A Manual of Dyeing*, 1893, 1:389; Chenciner, *Madder Red: A History of Luxury and Trade.*

88. Crookes, *A Practical Handbook of Dyeing and Calico-Printing*, 240, 254–8, 262, 314.

89. Kirby, Spring, and Higgitt, "The Technology of Eighteenth- and Nineteenth-Century Red Lake Pigments."

90. Archibald Orr Ewing & Co., "Turkey Red Dyeing Calculation Book AOE Lennoxbank."

91. Holme, "Sir William Henry Perkin: A Review of His Life, Work and Legacy"; Dépierre, *Traité de la teinture et de l'impression des matières colorantes artificielles. 2me partie : L'Alizarine artificielle et ses dérivés*, 2:1.

92. Dronsfield, Brown, and Cooksey, "Synthetic Alizarin- the Dye That Changed History."

93. Schunck, "XX. On the Colouring Matters of Madder"; Fieser, "The Discovery of Synthetic Alizarin."

94. Travis, "Between Broken Root and Artificial Alizarin: Textile Arts and Manufactures of Madder"; Perkin, "XVI. On Artificial Alizarin."

95. Fieser, "The Discovery of Synthetic Alizarin"; Graebe and Liebermann, "Ueber Alizarin und Anthracen"; Travis, *The Rainbow Makers: The Origins of the Synthetic Dyestuffs Industry in Western Europe*; Travis, "Between Broken Root and Artificial Alizarin: Textile Arts and Manufactures of Madder."

96. Travis, *The Rainbow Makers: The Origins of the Synthetic Dyestuffs Industry in Western Europe*; Gordon and Gregory, *Organic Chemistry in Colour*; Fieser, "The Discovery of Synthetic Alizarin"; Rowe, "The Life and Work of Sir William Henry Perkin."

97. Perkin, Coloring Matter, issued 1869.

98. Caro, Graebe, and Liebermann, Preparing Coloring Matters.

99. Travis, *The Rainbow Makers: The Origins of the Synthetic Dyestuffs Industry in Western Europe*, 180–83.

100. Perkin, "The History of Alizarin and Allied Colouring Matters, and Their Production from Coal Tar."

101. Perkin, Coloring Matter, issued 1870.

102. Bien, Stawitz, and Wunderlich, "Anthraquinone Dyes and Intermediates."

103. Perkin, "The History of Alizarin and Allied Colouring Matters, and Their Production from Coal Tar."
104. de Barry Barnett, *Anthracene and Anthraquinone*, 176–77.
105. Fieser, "The Discovery of Synthetic Alizarin."
106. Welham, "The Early History of the Synthetic Dye Industry."
107. Travis, *The Rainbow Makers: The Origins of the Synthetic Dyestuffs Industry in Western Europe*, 182–93.
108. Crookes, *A Practical Handbook of Dyeing and Calico-Printing*; Perkin, "The History of Alizarin and Allied Colouring Matters, and Their Production from Coal Tar"; Perkin, "Methods of Analysis Employed in the Manufacture of Alizarin."
109. Brandt, "Rapport présenté au nom du comité de chimie par M. Brandt, sur la valeur comparée de l'alizarine artificielle et de la garance."
110. Travis, *The Rainbow Makers: The Origins of the Synthetic Dyestuffs Industry in Western Europe*, 193–94.
111. Travis, "Between Broken Root and Artificial Alizarin: Textile Arts and Manufactures of Madder."
112. Lopez, "The Transition from Natural Madder to Synthetic Alizarine in the American Textile Industry, 1870–1890," 19.
113. Travis, *The Rainbow Makers: The Origins of the Synthetic Dyestuffs Industry in Western Europe*, 202.
114. Christie, "XII. Notes of Experiments on Artificial Alizarine"; Young, "XI.—On Artificial Alizarine"; Gordon and Gregory, *Organic Chemistry in Colour*, 11–12; Perkin, "XXXI. On the Formation of Anthrapurpurin."
115. Auerbach, *Anthracen*; Perkin, "XV. On Anthrapurpurin"; Dépierre, *Traité de la teinture et de l'impression des matières colorantes artificielles. 2me partie : L'Alizarine artificielle et ses dérivés*, 2:98–100; De Santis and Moresi, "Production of Alizarin Extracts from Rubia Tinctorum and Assessment of Their Dyeing Properties"; de Lalande, "Synthesis of Purpurin"; Knecht, Rawson, and Loewenthal, *A Manual of Dyeing*, 1893, 2:585.
116. Hornix, "From Process to Plant: Innovation in the Early Artificial Dye Industry"; Perkin, "Methods of Analysis Employed in the Manufacture of Alizarin"; Young, "XI.—On Artificial Alizarine"; Auerbach, *Anthracen*; Cain and Thorpe, *The Synthetic Dyestuffs and the Intermediate Products from Which They Are Derived*; Lopez, "The Transition from Natural Madder to Synthetic Alizarine in the American Textile Industry, 1870–1890," 19; Dépierre, *Traité de la teinture et de l'impression des matières colorantes artificielles. 2me partie : L'Alizarine artificielle et ses dérivés*, 2:68–72.
117. Rowe, "The Life and Work of Sir William Henry Perkin"; Perkin, "The History of Alizarin and Allied Colouring Matters, and Their Production from Coal Tar"; Verbong, "Turksrood."

Chapter 3

1. Venkataraman, *The Chemistry of Synthetic Dyes*, 1:279.
2. Henry, "Considerations Relative to the Nature of Wool, Silk, and Cotton, as Objects of the Art of Dying; on the Various Preparations, and Mordants, Requisite for These Different Substances; and on the Nature and Properties of Colouring Matter. Together with Some."
3. Chaptal, *L'art de la teinture du coton en rouge*.
4. Crookes, *A Practical Handbook of Dyeing and Calico-Printing*, 325.
5. Sansone, "Alizarin-Red and Turkey-Red Dyeing and Printing on Cotton."
6. Carruthers, "A New Process of Dyeing Turkey Red."

7. Kendall, "Preparation of Alizarine Assistant and Its Action in Turkey-Red Dyeing."

8. Karadag and Dolen, "Re-Examination of Turkey Red."

9. Mazeas, "Recherches sur la cause physique de l'adhérence de la couleur rouge aux Toiles peintes qui nous viennent des cotes de Malabar & de Coromandel"; Berthollet, *Elements of the Art of Dyeing.*

10. Knecht, Rawson, and Loewenthal, *A Manual of Dyeing*, 1893, 1:16–24.

11. Chettra, "Microscopy and Surface Chemical Investigations of Dyed Cellulose Textiles," 12.

12. Hummel, *The Dyeing of Textile Fabrics*, 20.

13. Chateau, "Critical and Historical Notes Concerning the Production of Adrianople or Turkey Red, and the Theory of This Colour No. 7"; Jenny, "Mémoire sur la fabrication du rouge d'Andrinople, présenté par M. Jenny, et traduit de l'allemand par M. Rosenstiehl."

14. Chateau, "Critical and Historical Notes Concerning the Production of Adrianople or Turkey Red, and the Theory of This Colour No. 10."

15. Brandt et al., "Sulfoléates et rouge turc par le procédé rapide, priorités"; Le Pileur d'Apligny, *L'art de la teinture des fils et étoffes de coton*, 143; Hellot, Macquer, and Le Pileur d'Apligny, *The Art of Dyeing Wool, Silk, and Cotton*, 408; Schaefer, "The History of Turkey Red Dyeing"; Pallas, "I. The Genuine Oriental Process for Giving to Cotton Yarn or Stuff the Fast or Ingrained Colour, Known by the Name of Turkey Red, as Practised at Astracan"; Cunningham et al., "Hanging by a Thread: Natural Metallic Mordant Processes in Traditional Indonesian Textiles"; Crookes, *A Practical Handbook of Dyeing and Calico-Printing*, 321; Vitalis, *Cours élémentaire de teinture sur laine, soie, lin, chanvre et coton, et sur l'art d'imprimer les toiles*, 118; Anderson and Lowe, "The Composition of Flaxseed Mucilage"; Chateau, "Critical and Historical Notes Concerning the Production of Adrianople or Turkey Red, and the Theory of This Colour No. 7," 31–2; Jenny, "Mémoire sur la fabrication du rouge d'Andrinople, présenté par M. Jenny, et traduit de l'allemand par M. Rosenstiehl," 757; "Dyeing Turkey Red."

16. Haussmann, "XLIII. Observations on Maddering; Together with a Simple and Certain Process for Obtaining, with Great Beauty and Fixity, That Colour Known under the Name of the Turkey or Adrianople Red."

17. Le Pileur d'Apligny, *Essai sur les moyens de perfectionner l'art de la teinture, et observations sur quelques matières qui y sont propres*; Berthollet, *Elements of the Art of Dyeing*, 145–56; Crookes, *Dyeing and Tissue Printing*, 76; Hellot, Macquer, and Le Pileur d'Apligny, *The Art of Dyeing Wool, Silk, and Cotton*; Lalande, "L'art de faire le maroquin"; Chateau, "Critical and Historical Notes Concerning the Production of Adrianople or Turkey Red, and the Theory of This Colour No. 7"; Le Pileur d'Apligny, *L'art de la teinture des fils et étoffes de coton*, 136–37; Haussmann, "XXXII. Observations on Maddering; Together with a Simple and Certain Process for Obtaining, with Great Beauty and Fixity, That Colour Known under the Name of the Turkey or Adrianople Red"; Vitalis, "Mémoire sur la nature la fiente de mouton, et sur son usage dans la teinture du coton en rouge di des Indes, ou d'Andrinople," 35; Persoz, *Traité théorique et pratique de l'impression des tissus*, 1846, 3:187–88; Vitalis, *Cours élémentaire de teinture sur laine, soie, lin, chanvre et coton, et sur l'art d'imprimer les toiles*; Ure, *A Dictionary of Arts, Manufactures, and Mines*; Schaefer, "The History of Turkey Red Dyeing."

18. Rauch, *John Rauch's Receipts on Dyeing, in a Series of Letters to a Friend. Containing Correct and Exact Copies of All His Best Receipts on Dyeing Cotton and Woollen Goods; Obtained and Improved by Him, during Twelve Years Practice, at Different Manufactories in Switzerland, France, Germany and America. Also, a true Description of His Invented Substitute for Woad, Being a Cheap and Preferable*

Material, and the Produce of this Country; Chateau, "Critical and Historical Notes Concerning the
Production of Adrianople or Turkey Red, and the Theory of This Colour No. 4"; Sansone, "Alizarin-
Red and Turkey-Red Dyeing and Printing on Cotton"; Vitalis, *Cours élémentaire de teinture sur laine,
soie, lin, chanvre et coton, et sur l'art d'imprimer les toiles*; Carruthers, "A New Process of Dyeing
Turkey Red"; Haller, "The Chemistry and Technique of Turkey Red Dyeing"; Schaefer, "The History
of Turkey Red Dyeing"; Hellot, Macquer, and Le Pileur d'Apligny, *The Art of Dyeing Wool, Silk, and
Cotton*; "V. Account of the Process Followed by M. Pierre Jaques Papillon for Dyeing Turkey Red";
Chateau, "Critical and Historical Notes Concerning the Production of Adrianople or Turkey Red,
and the Theory of This Colour No. 6"; Pallas, "I. The Genuine Oriental Process for Giving to Cotton
Yarn or Stuff the Fast or Ingrained Colour, Known by the Name of Turkey Red, as Practised at
Astracan"; Henry, "Considerations Relative to the Nature of Wool, Silk, and Cotton, as Objects of the
Art of Dying; on the Various Preparations, and Mordants, Requisite for These Different Substances;
and on the Nature and Properties of Colouring Matter. Together with Some"; O'Neill, "The Printing
and Dyeing of Calico, Silk, and Woollen Fabrics"; Haussmann, "XXXII. Observations on Maddering;
Together with a Simple and Certain Process for Obtaining, with Great Beauty and Fixity, That
Colour Known under the Name of the Turkey or Adrianople Red"; Le Pileur d'Apligny, *L'art de la
teinture des fils et étoffes de coton*.

19. Jenny, "Mémoire sur la fabrication du rouge d'Andrinople, présenté par M. Jenny, et traduit de
l'allemand par M. Rosenstiehl," 759.
20. Pallas, "I. The Genuine Oriental Process for Giving to Cotton Yarn or Stuff the Fast or Ingrained
Colour, Known by the Name of Turkey Red, as Practised at Astracan"; Verbong, "Turksrood," 274.
21. Chateau, "Critical and Historical Notes Concerning the Production of Adrianople or Turkey Red,
and the Theory of This Colour No. 6," 384.
22. Persoz, *Traité théorique et pratique de l'impression des tissus*, 1846, 3:211; Crookes, *A Practical
Handbook of Dyeing and Calico-Printing*, 325.
23. Henry, "Considerations Relative to the Nature of Wool, Silk, and Cotton, as Objects of the Art of
Dying; on the Various Preparations, and Mordants, Requisite for These Different Substances; and
on the Nature and Properties of Colouring Matter. Together with Some."
24. "V. Account of the Process Followed by M. Pierre Jaques Papillon for Dyeing Turkey Red."
25. Rauch, *John Rauch's Receipts on Dyeing, in a Series of Letters to a Friend. Containing Correct and
Exact Copies of All His Best Receipts on Dyeing Cotton and Woollen Goods; Obtained and Improved
by Him, during Twelve Years Practice, at Different Manufactories in Switzerland, France, Germany
and America. Also, a true Description of His Invented Substitute for Woad, Being a Cheap and
Preferable Material, and the Produce of this Country*; Factories Inquiry Commission,
"Supplementary Report of the Central Board of His Majesty's Commissioners Appointed to
Collect Information in the Manufacturing Districts, as to the Employment of Children in
Factories, and as to the Propriety and Means of Curtailing the Hours of Their Lab."
26. Chateau, "Critical and Historical Notes Concerning the Production of Adrianople or Turkey Red,
and the Theory of This Colour No. 3," 174.
27. Chateau, "Critical and Historical Notes Concerning the Production of Adrianople or Turkey Red,
and the Theory of This Colour No. 6," 384.
28. Hellot, Macquer, and Le Pileur d'Apligny, *The Art of Dyeing Wool, Silk, and Cotton*, 407–8.
29. Hummel, *The Dyeing of Textile Fabrics*, 384.
30. Dumas, *Traité de chimie, appliquée aux arts*; Henry, "Considerations Relative to the Nature of
Wool, Silk, and Cotton, as Objects of the Art of Dying; on the Various Preparations, and

Mordants, Requisite for These Different Substances; and on the Nature and Properties of Colouring Matter. Together with Some"; "V. Account of the Process Followed by M. Pierre Jaques Papillon for Dyeing Turkey Red."

31. Persoz, *Traité théorique et pratique de l'impression des tissus*, 1846.

32. Dépierre, *Traité de la teinture et de l'impression des matières colorantes artificielles. 2me partie : L'Alizarine artificielle et ses dérivés*, 2:397.

33. Thomson, "I. New Theory of Dyeing Turkey Red."

34. Crookes, *A Practical Handbook of Dyeing and Calico-Printing*, 324.

35. Chateau, "Critical and Historical Notes Concerning the Production of Adrianople or Turkey Red, and the Theory of This Colour No. 9."

36. Parks, "The Chemistry of Turkey-Red Dyeing"; Schutzenberger, *Traite des Matieres Colorantes*.

37. Dumas, *Traité de chimie, appliquée aux arts*; Thomson, "I. New Theory of Dyeing Turkey Red."

38. Henry, "Considerations Relative to the Nature of Wool, Silk, and Cotton, as Objects of the Art of Dying; on the Various Preparations, and Mordants, Requisite for These Different Substances; and on the Nature and Properties of Colouring Matter. Together with Some," 381; Knecht, Rawson, and Loewenthal, *A Manual of Dyeing*, 1893, 2:593; Persoz, *Traité théorique et pratique de l'impression des tissus*, 1846.

39. Knecht, Rawson, and Loewenthal, *A Manual of Dyeing*, 1893, 2:589.

40. Kiel, "Metaalcomplexen van alizarinerood"; Kiel and Heertjes, "Metal Complexes of Alizarin V. Investigations of Alizarin–Dyed Cotton Fabrics."

41. Wertz et al., "Characterisation of Oil and Aluminium Complex on Replica and Historical 19th c. Turkey Red Textiles by Non-Destructive Diffuse Reflectance FTIR Spectroscopy."

42. Brennan, Tannahill, and Percy, "Turkey Red Process for Cotton Yarn"; Tannahill, "Turkey Red Dyeing"; Collin, "Turkey Red Process for Cotton Yarn"; Archibald Orr Ewing & Co., "Turkey Red Dyeing Calculation Book AOE Lennoxbank."

43. "V. Account of the Process Followed by M. Pierre Jaques Papillon for Dyeing Turkey Red"; Vitalis, *Cours élémentaire de teinture sur laine, soie, lin, chanvre et coton, et sur l'art d'imprimer les toiles*, 236–38; Henry, "Considerations Relative to the Nature of Wool, Silk, and Cotton, as Objects of the Art of Dying; on the Various Preparations, and Mordants, Requisite for These Different Substances; and on the Nature and Properties of Colouring Matter. Together with Some," 383; Vitalis, *Essai sur l'origine et les progrès de l'art de la teinture en France, et particulièrement de l'art de teindre le coton en rouge dit des Indes. Lu à la Société de commerce de Rouen*, 21; Clements and Sadler, *The New Handmaid to Arts Sciences Agriculture*.

44. Knecht, Rawson, and Loewenthal, *A Manual of Dyeing*, 1893, 2:591–94.

45. Vitalis, *Cours élémentaire de teinture sur laine, soie, lin, chanvre et coton, et sur l'art d'imprimer les toiles*.

46. Chateau, "Critical and Historical Notes Concerning the Production of Adrianople or Turkey Red, and the Theory of This Colour No. 10"; MacFarlane, *A Practical Treatise on Dyeing and Calico-Printing*; Hellot, Macquer, and Le Pileur d'Apligny, *The Art of Dyeing Wool, Silk, and Cotton*.

47. Chateau, "Critical and Historical Notes Concerning the Production of Adrianople or Turkey Red, and the Theory of This Colour No. 4," 222.

48. Cain and Thorpe, *The Synthetic Dyestuffs and the Intermediate Products from Which They Are Derived*; Berthollet, *Elements of the Art of Dyeing*.

49. Mohanty, Chandramouli, and Naik, *Natural Dyeing Processes of India*; Knecht, Rawson, and Loewenthal, *A Manual of Dyeing*, 1893, 1:174.

50. Dumas, *Traité de chimie, appliquée aux arts.*

51. Hummel, *The Dyeing of Textile Fabrics*; Berthollet, *Elements of the Art of Dyeing.*

52. Mohanty, Chandramouli, and Naik, *Natural Dyeing Processes of India*, 12.

53. Vitalis, "Mémoire sur la nature la fiente de mouton, et sur son usage dans la teinture du coton en rouge di des Indes, ou d'Andrinople," 155; Knecht, Rawson, and Loewenthal, *A Manual of Dyeing*, 1893, 1:211–12; Henry, "Considerations Relative to the Nature of Wool, Silk, and Cotton, as Objects of the Art of Dying; on the Various Preparations, and Mordants, Requisite for These Different Substances; and on the Nature and Properties of Colouring Matter. Together with Some"; Berthoud, *Les indiennes neuchâteloises*, 195–96.

54. Koller, *The Utilization of Waste Products.*

55. Knecht, Rawson, and Loewenthal, *A Manual of Dyeing*, 1893, 1:211.

56. Tannahill, "Turkey Red Dyeing"; Archibald Orr Ewing & Co., "Turkey Red Dyeing Calculation Book AOE Lennoxbank."

57. Crookes, *A Practical Handbook of Dyeing and Calico-Printing*, 294–97.

58. Clow and Clow, "The Chemical Revolution: A Contribution to Social Technology," 202.

59. Hummel, *The Dyeing of Textile Fabrics*, 227–32; Espinosa-Jiménez et al., "The Adsorption of Tannic Acid on Hydrophilic Cotton and Its Effect on the Electrokinetic Properties of This Cellulose Fibre in a Cationic Dye Solution"; Berthollet, *Elements of the Art of Dyeing*; Fereday, *Natural Dyes.*

60. Markley, *Fatty Acids: Their Chemistry, Properties, Production, and Uses.*

61. Burton and Robertshaw, *Sulphated Oils and Allied Products*, 1.

62. "V. Account of the Process Followed by M. Pierre Jaques Papillon for Dyeing Turkey Red."

63. Radcliffe and Medofski, "The Sulphonation of Fixed Oils."

64. Chateau, "Critical and Historical Notes Concerning the Production of Adrianople or Turkey Red, and the Theory of This Colour No. 6," 391.

65. Hurst, *Textile Soaps and Oils.*

66. Burton and Robertshaw, *Sulphated Oils and Allied Products*, 2.

67. Naughton, "Production, Chemistry, and Commercial Applications of Various Chemicals from Castor Oil."

68. Markley, *Fatty Acids: Their Chemistry, Properties, Production, and Uses.*

69. Trask, "Sulfonation and Sulfation of Oils."

70. Leigh, "On the Estimation of Alizarin in Dyed Cotton Fabrics, and on an Attempt to Ascertain the Composition of Turkey-Red and Other Alizarin Lakes."

71. Radcliffe and Medofski, "The Sulphonation of Fixed Oils"; Brandt et al., "Sulfoléates et rouge turc par le procédé rapide, priorités."

72. Knecht, Rawson, and Loewenthal, *A Manual of Dyeing*, 1893, 1:161; Tannahill, "Turkey Red Dyeing."

73. Storey, *The Thames and Hudson Manual of Dyes and Fabrics*, 89; Gunstone and Padley, *Lipid Technologies and Applications.*

74. Straugh, "Recipe from Mr Straugh"; Hurst, *Textile Soaps and Oils*; Radcliffe and Medofski, "The Sulphonation of Fixed Oils."

75. Trask, "Sulfonation and Sulfation of Oils"; Hurst, "Recent Progress in Dyeing"; Burton and Robertshaw, *Sulphated Oils and Allied Products*; Markley, *Fatty Acids: Their Chemistry, Properties, Production, and Uses.*

76. Naughton, "Production, Chemistry, and Commercial Applications of Various Chemicals from Castor Oil."

77. Burton and Robertshaw, *Sulphated Oils and Allied Products*; Wilson, "Turkey-Red Oil Part II."; Brandt et al., "Sulfoléates et rouge turc par le procédé rapide, priorités"; Hurst, *Textile Soaps and Oils*; Knecht, Rawson, and Loewenthal, *A Manual of Dyeing*, 1893.

78. Kendall, "Preparation of Alizarine Assistant and Its Action in Turkey-Red Dyeing."

79. Wertz, France, and Quye, "Spectroscopic Analysis of Turkey Red Oil Samples as a Basis for Understanding Historical Dyed Textiles."

80. Sansone, "Alizarin-Red and Turkey-Red Dyeing and Printing on Cotton"; Dunbar, "Modern Developments in Textile Chemicals for Dyeing and Finishing"; Achaya, "Chemical Derivatives of Castor Oil"; Gunstone and Padley, *Lipid Technologies and Applications*; Trotman, *Dyeing and Chemical Technology of Textile Fibres*, 202; Radcliffe and Medofski, "The Sulphonation of Fixed Oils"; Ahmad and Singh, "Surface Active Properties of Sulfonated Isoricinoleic Acid."

81. Carruthers, "A New Process of Dyeing Turkey Red."

82. Hummel, *The Dyeing of Textile Fabrics*, 443–5.

83. de Niederhäusern, "Rapport sur le travail de M. F. Driessen: 'Etude sur le rouge turc, ancien procédé', et sur le contenu de deux plis cachetés Nos. 700 et 1276, déposés par le même auteur."

84. Verbong, "Turksrood."

85. Peel, "Turkey Red Dyeing in Scotland Its Heyday and Decline."

86. Tannahill, "Turkey Red Dyeing."

87. Verbong, "Technische innovaties in de katoendrukkerij en -ververij in Nederland 1835–1920," 279.

88. Travis, "Between Broken Root and Artificial Alizarin: Textile Arts and Manufactures of Madder"; Jacqué et al., *Andrinople: Le rouge magnifique*; Nieto-Galan, *Colouring Textiles : A History of Natural Dyestuffs in Industrial Europe*.

89. Hummel, *The Dyeing of Textile Fabrics*, 438–42.

90. Jacqué et al., *Andrinople: Le rouge magnifique*; Travis, "Between Broken Root and Artificial Alizarin: Textile Arts and Manufactures of Madder."

91. Jacqué et al., *Andrinople: Le rouge magnifique*.

92. Persoz, *Traité théorique et pratique de l'impression des tissus*, 1846, 3:453.

93. Driessen, "Étude sur le rouge turc, ancien procédé," 164.

94. Henry, "Considerations Relative to the Nature of Wool, Silk, and Cotton, as Objects of the Art of Dying; on the Various Preparations, and Mordants, Requisite for These Different Substances; and on the Nature and Properties of Colouring Matter. Together with Some"; Schaefer, "The History of Turkey Red Dyeing"; Berthollet, *Elements of the Art of Dyeing*.

95. Berthollet.

96. Fairlie, "Dyestuffs in the Eighteenth Century."

97. Vitalis, *Cours élémentaire de teinture sur laine, soie, lin, chanvre et coton, et sur l'art d'imprimer les toiles*; Hummel, *The Dyeing of Textile Fabrics*; Knecht, Rawson, and Loewenthal, *A Manual of Dyeing*, 1893.

98. Crookes, *Dyeing and Tissue Printing*.

99. Cain and Thorpe, *The Synthetic Dyestuffs and the Intermediate Products from Which They Are Derived*.

100. Hummel, *The Dyeing of Textile Fabrics*, 432–3.

101. Knecht, Rawson, and Loewenthal, *A Manual of Dyeing*, 1893, 1:234; Berthollet, *Elements of the Art of Dyeing*.

102. Lynde, *The Domestic Dyer, or Philosophy of Fast Colours: Being a Compilation from the Most Approved American and European Authors*; Dépierre, *Traité de la teinture et de l'impression des matières colorantes artificielles. 2me partie : L'Alizarine artificielle et ses dérivés*, 2:412–14.

103. Persoz, *Traité théorique et pratique de l'impression des tissus*, 1846, 3:210; Dépierre, *Traité de la teinture et de l'impression des matières colorantes artificielles. 2me partie : L'Alizarine artificielle et ses dérivés*, 2:412.

104. Chateau, "Critical and Historical Notes Concerning the Production of Adrianople or Turkey Red, and the Theory of This Colour No. 9," 139; Bossert, "The Metallic Soaps"; Jenny, "Mémoire sur la fabrication du rouge d'Andrinople, présenté par M. Jenny, et traduit de l'allemand par M. Rosenstiehl," 821; Wertz et al., "Characterisation of Oil and Aluminium Complex on Replica and Historical 19th c. Turkey Red Textiles by Non-Destructive Diffuse Reflectance FTIR Spectroscopy"; Kirby, Spring, and Higgitt, "The Technology of Eighteenth- and Nineteenth-Century Red Lake Pigments"; Kiel and Heertjes, "Metal Complexes of Alizarin V. Investigations of Alizarin–Dyed Cotton Fabrics."

105. Hummel, *The Dyeing of Textile Fabrics*, 433–4.

106. Chateau, "Critical and Historical Notes Concerning the Production of Adrianople or Turkey Red, and the Theory of This Colour No. 4," 219–21; Ure, *A Dictionary of Arts, Manufactures, and Mines*.

107. Sansone, "Alizarin-Red and Turkey-Red Dyeing and Printing on Cotton."

108. Baker, "East Indian Hand-Painted Calicoes of the Seventeenth and Eighteenth Centuries, and Their Influence on the Tinctorial Arts of Europe"; Persoz, *Traité théorique et pratique de l'impression des tissus*, 1846, 3:182.

109. Ure, *A Dictionary of Arts, Manufactures, and Mines*.

110. Haussmann, "XXXII. Observations on Maddering; Together with a Simple and Certain Process for Obtaining, with Great Beauty and Fixity, That Colour Known under the Name of the Turkey or Adrianople Red."

111. Crookes, *Dyeing and Tissue Printing*.

112. Le Pileur d'Apligny, *L'art de la teinture des fils et étoffes de coton*; Cunningham et al., "Hanging by a Thread: Natural Metallic Mordant Processes in Traditional Indonesian Textiles."

113. Vitalis, *Cours élémentaire de teinture sur laine, soie, lin, chanvre et coton, et sur l'art d'imprimer les toiles*.

114. Archibald Orr Ewing & Co., "Turkey Red Dyeing Calculation Book AOE Lennoxbank."

115. Verbong, "Technische innovaties in de katoendrukkerij en -ververij in Nederland 1835-1920," 53–66.

116. Schaefer, "The Cultivation of Madder."

117. Richter, "LXXXII. Vital Staining of Bones with Madder."

118. Müller-Jacobs, "Turkey-Red from Alizarine."

119. Meyer et al., "In-Situ Spectroscopic Analysis of the Traditional Dyeing Pigment Turkey Red inside Textile Matrix"; Zhuang et al., "New Insights into the Structure and Degradation of Alizarin Lake Pigments: Input of the Surface Study Approach."

120. Bergerhoff and Wunderlich, "Crystal Structure of Di-μ-Oxo-Bis(Bis(1,2,4-Trihydroxyanthrachinonato)Aluminium-(Aquabis(Dimethylformamide)Calcium)), $Al_2O_4(C_{14}H_6O_5)_4Ca_2(HCON(CH_3)_2)_4(H_2O)_2$, (Purpurine Complex)"; Wunderlich and Bergerhoff, "Crystal Structure of Di-μ-Oxo-Bis(Bis(1,2-Dihydroxyanthrachinonato)Aluminium-(Aquabis(Dimethylformamide)Calcium)), $Al_2O_4(C_{14}H_6O_4)_4Ca_2(HCON(CH_3)_2)_4(H_2O)_2$, (Alizarine Complex)."

121. Soubayrol et al., "Spectrométrie de masse et chromatographie liquides des laques complexes des teintures à l'alizarine"; Soubayrol, "Preparation et etude structurale des complexes formes entre l'aluminium et l'alizarine. Importance de la nature du solvant et de la base utilises surle degre de

condensation de l'aluminium et l'agencement moleculaire"; Soubayrol, Dana, and Man, "Aluminium-27 Solid-State NMR Study of Aluminium Coordination Complexes of Alizarin."

122. Delamare, Monasse, and Garcia, "The Role of Aluminium as a Mordant for Cellulose Dyeing with Alizarin: A Numerical Approach."

123. Wertz, "Turkey Red Dyeing in Late-19th Century Glasgow: Interpreting the Historical Process through Re-Creation and Chemical Analysis for Heritage Research and Conservation."

124. Blackburn, "Natural Dyes in Madder (Rubia Spp.) and Their Extraction and Analysis in Historical Textiles"; Kiel, "Metaalcomplexen van alizarinerood"; de Barry Barnett, *Anthracene and Anthraquinone*; Kiel and Heertjes, "I. The Metal Complexes of Alizarin Structure of the Calcium-Aluminium Lake of Alizarin"; Sansone, "Alizarin-Red and Turkey-Red Dyeing and Printing on Cotton"; Fierz-David and Rutishauser, "Composition and Constitution of Turkey Red."

125. Chateau, "Critical and Historical Notes Concerning the Production of Adrianople or Turkey Red, and the Theory of This Colour No. 11," 265; Thomson, "I. New Theory of Dyeing Turkey Red"; Hummel, *The Dyeing of Textile Fabrics*; Chateau, "Critical and Historical Notes Concerning the Production of Adrianople or Turkey Red, and the Theory of This Colour No. 9"; Pallas, "II. Process for Dyeing the Adrianople or Turkey Red, as Practised at Astracan. Being a Supplement to His Former Publications on That Art"; Schaefer, "The History of Turkey Red Dyeing"; Dépierre, *Traité de la teinture et de l'impression des matières colorantes artificielles. 2me partie : L'Alizarine artificielle et ses dérivés*, 2:396.

126. "The Manufacture of Articles of Commerce from Blood."

127. Bremner, *The Industries of Scotland: Their Rise, Progress, and Present Condition.*

128. Persoz, *Traité théorique et pratique de l'impression des tissus*, 1846, 1:316; Crookes, *A Practical Handbook of Dyeing and Calico-Printing*; "The Manufacture of Articles of Commerce from Blood."

129. Koller, *The Utilization of Waste Products.*

130. Tannahill, "Turkey Red Dyeing."

131. Hummel, *The Dyeing of Textile Fabrics*; Pallas, "I. The Genuine Oriental Process for Giving to Cotton Yarn or Stuff the Fast or Ingrained Colour, Known by the Name of Turkey Red, as Practised at Astracan"; "V. Account of the Process Followed by M. Pierre Jaques Papillon for Dyeing Turkey Red"; Crookes, *A Practical Handbook of Dyeing and Calico-Printing*; Haussmann, "XXXII. Observations on Maddering; Together with a Simple and Certain Process for Obtaining, with Great Beauty and Fixity, That Colour Known under the Name of the Turkey or Adrianople Red"; Henry, "Considerations Relative to the Nature of Wool, Silk, and Cotton, as Objects of the Art of Dying; on the Various Preparations, and Mordants, Requisite for These Different Substances; and on the Nature and Properties of Colouring Matter. Together with Some."

132. Chateau, "Critical and Historical Notes Concerning the Production of Adrianople or Turkey Red, and the Theory of This Colour No. 3"; Crookes, *A Practical Handbook of Dyeing and Calico-Printing*; Dépierre, *Traité de la teinture et de l'impression des matières colorantes artificielles. 2me partie : L'Alizarine artificielle et ses dérivés*, 2:397; Hummel, *The Dyeing of Textile Fabrics*, 436.

133. Schaefer, "The History of Turkey Red Dyeing."

Chapter 4

1.   Hinckley, "Everyday Actualities. No. II."

2.   Tarrant, "The Turkey Red Dyeing Industry in the Vale of Leven."

3.   Storey, "Turkey Red Prints."
4.   Bremner, *The Industries of Scotland: Their Rise, Progress, and Present Condition*, 302–3.
5.   Hinckley, "Everyday Actualities. No. II," 7; Bremner, *The Industries of Scotland: Their Rise, Progress, and Present Condition*; Robinson, *A History of Printed Textiles*, 24–6; Kusamitsu, "British Industrialization and Design before the Great Exhibition."
6.   Crookes, *A Practical Handbook of Dyeing and Calico-Printing*, 317.
7.   Trotman, *Dyeing and Chemical Technology of Textile Fibres*.
8.   Persoz, *Traité théorique et pratique de l'impression des tissus*, 1846, 3:234–37.
9.   Peel, "Turkey Red Dyeing in Scotland Its Heyday and Decline."
10. Ure, "Description of the Great Bandana Gallery in the Turkey-Red Factory of Messrs. Monteith & Co. at Glasgow."
11. "On the Process for Discharging Turkey-Red."
12. H., "On the Process for Discharging Turkey-Red."
13. "Notices to Correspondents. War on Turkey-Red."
14. "Discharging of Turkey-Red."
15. Harvey, "Statement of Claims on the Invention of the Process for Discharging Turkey-Red"; "Notices to Correspondents."
16. "Notices to Correspondents: Bandana Controversy."
17. Miller, "Statement Relative to the Discharging Process of Turkey Red, by Means of Presses."
18. Campbell, "Mr. Campbell's Proof of Claims."
19. Miller, "Statement of Claims to the Invention of the Process for Discharging Turkey-Red."
20. Campbell, "Statement of Claims to the Invention of the Process for Discharging Turkey-Red."
21. Miller, "Reply to Mr. David Campbell's Statement of Claims to the Invention of the Process for Discharging Turkey Red. Inserted in *The Glasgow Mechanics' Magazine*, No. LXXXIX, Sep. 3d 1825."
22. Millar, "Mr. Millar's Statement of Claims, &c."
23. Arthur et al., *Seeing Red: Scotland's Exotic Textile Heritage*, 7.
24. Bremner, *The Industries of Scotland: Their Rise, Progress, and Present Condition*, 299; Persoz, *Traité théorique et pratique de l'impression des tissus*, 1846, 3:234–7.
25. "Specification of the Patent Granted to James Thomson, of Primrose Hill, near Clithero, in the County of Lancaster, Calico Printer; for a Method of Producing Patterns on Cloth Peviously Dyed Turkey Red, and Made of Cotton or Linen, or Both."
26. Crookes, *Dyeing and Tissue Printing*.
27. Hargreaves, *Messrs. Hargreaves' Calico Print Works at Accrington, and Recollections of Broad Oak*, 7–8.
28. Storey, "Turkey Red Prints"; "On the Process for Discharging Turkey-Red"; Henry, "Considerations Relative to the Nature of Wool, Silk, and Cotton, as Objects of the Art of Dying; on the Various Preparations, and Mordants, Requisite for These Different Substances; and on the Nature and Properties of Colouring Matter. Together with Some."
29. Storey, "Turkey Red Prints."
30. Persoz, *Traité théorique et pratique de l'impression des tissus*, 1846.
31. Storey, "Turkey Red Prints."
32. Schaefer, "The History of Turkey Red Dyeing."
33. Hargreaves, *Messrs. Hargreaves' Calico Print Works at Accrington, and Recollections of Broad Oak*, 8.
34. Greene, *Wearable Prints, 1760–1860: History, Materials, and Mechanics*.
35. Storey, "Turkey Red Prints"; Schaefer, "The History of Turkey Red Dyeing"; Crookes, *A Practical Handbook of Dyeing and Calico-Printing*.

36. Crookes, *Dyeing and Tissue Printing*.
37. Crookes, *A Practical Handbook of Dyeing and Calico-Printing*, 156.
38. Storey, "Turkey Red Prints."
39. Wertz et al., "Characterisation of Oil and Aluminium Complex on Replica and Historical 19th c. Turkey Red Textiles by Non-Destructive Diffuse Reflectance FTIR Spectroscopy."
40. Wertz, Quye, and France, "Turkey Red Prints: Identification of Lead Chromate, Prussian Blue and Logwood on Turkey Red Calico."
41. Clark, "The Design and Designing of Lancashire Printed Calicoes during the First Half of the 19th Century."
42. Clark.
43. Kusamitsu, "British Industrialization and Design before the Great Exhibition."
44. Greysmith, "Patterns, Piracy and Protection in the Textile Printing Industry 1787–1850."
45. Tuckett and Nenadic, "Colouring the Nation: A New In-Depth Study of the Turkey Red Pattern Books in the National Museums Scotland."
46. Greysmith, "Patterns, Piracy and Protection in the Textile Printing Industry 1787–1850"; Eastop, "New Ways of Engaging with Historic Textiles: Interactive Images Online"; Arthur et al., *Seeing Red: Scotland's Exotic Textile Heritage*, 22–3; Kusamitsu, "British Industrialization and Design before the Great Exhibition"; Lopez, "The Transition from Natural Madder to Synthetic Alizarine in the American Textile Industry, 1870–1890."
47. Bremner, *The Industries of Scotland: Their Rise, Progress, and Present Condition*, 302.
48. Fryer, *Behind the Vale*.
49. Arthur et al., *Seeing Red: Scotland's Exotic Textile Heritage*, 22.
50. Clark, "The Design and Designing of Lancashire Printed Calicoes during the First Half of the 19th Century."
51. Nenadic, "Selling Printed Cottons in Mid-Nineteenth-Century India: John Matheson of Glasgow and Scottish Turkey Red."
52. İnalcik, "When and How British Cotton Goods Invaded the Levant Markets."
53. Schaefer, "The History of Turkey Red Dyeing"; Sandberg, *The Red Dyes*, 146.
54. Persoz, *Traité théorique et pratique de l'impression des tissus*, 1846, 4:428; Robinson, *A History of Printed Textiles*, 115.
55. Nenadic, "Selling Printed Cottons in Mid-Nineteenth-Century India: John Matheson of Glasgow and Scottish Turkey Red."
56. Nenadic.
57. Arthur et al., *Seeing Red: Scotland's Exotic Textile Heritage*, 22.
58. Clark, "The Design and Designing of Lancashire Printed Calicoes during the First Half of the 19th Century."
59. Swain, "Turkey Red"; Nenadic, "Selling Printed Cottons in Mid-Nineteenth-Century India: John Matheson of Glasgow and Scottish Turkey Red"; Tuckett and Nenadic, "Colouring the Nation: A New In-Depth Study of the Turkey Red Pattern Books in the National Museums Scotland."
60. Jacqué et al., *Andrinople: Le rouge magnifique*, 68–69.
61. Tuckett and Nenadic, "Colouring the Nation: A New In-Depth Study of the Turkey Red Pattern Books in the National Museums Scotland."
62. Storey, "Turkey Red Prints."
63. O'Neill, "The Printing and Dyeing of Calico, Silk, and Woollen Fabrics," 358.

Chapter 5

1. Hanna, *Ottoman Egypt and the Emergence of the Modern World: 1500-1800*, 102–3.
2. Clow and Clow, "The Chemical Revolution: A Contribution to Social Technology," 214–15; Faroqhi, "Ottoman Cotton Textiles, 1500s to 1800: The Story of a Success That Did Not Last."
3. Scoggin, "A Piece of Red Calico."
4. Biggam, "Rags to Riches."
5. Peel, "Turkey Red Dyeing in Scotland Its Heyday and Decline."
6. Nenadic, "Selling Printed Cottons in Mid-Nineteenth-Century India: John Matheson of Glasgow and Scottish Turkey Red."
7. Fox, "Presidential Address: Science, Industry, and the Social Order in Mulhouse, 1798-1871."
8. Travis, "Between Broken Root and Artificial Alizarin: Textile Arts and Manufactures of Madder"; Schaefer, "The History of Turkey Red Dyeing."
9. Johnston, "The Secret of Turkey Red Technology Transfer with a Scottish Connection."
10. Vitalis, *Essai sur l'origine et les progrès de l'art de la teinture en France, et particulièrement de l'art de teindre le coton en rouge dit des Indes. Lu à la Société de commerce de Rouen*, 38–9; Dépierre, *Traité de la teinture et de l'impression des matières colorantes artificielles. 2me partie : L'Alizarine artificielle et ses dérivés*, 2:368.
11. Schmitt, "Relations between England and the Mulhouse Textile Industry in the Nineteenth Century."
12. Schaefer, "The History of Turkey Red Dyeing."
13. Jacqué et al., *Andrinople: Le rouge magnifique*, 14–15.
14. Smith, *The Emergence of Modern Business Enterprise in France, 1800-1930*; Persoz, *Traité théorique et pratique de l'impression des tissus*; Dépierre, *Traité de la teinture et de l'impression des matières colorantes artificielles. 2me partie : L'Alizarine artificielle et ses dérivés*, 2:392; Adams, "Four Aspects of Turkey Red: Turkey Red in Quilts and Clothing."
15. Persoz, *Traité théorique et pratique de l'impression des tissus*.
16. Clark, "The Design and Designing of Lancashire Printed Calicoes during the First Half of the 19th Century."
17. Aikin, *A Description of the Country from Thirty to Forty Miles Round Manchester*.
18. Cliffe, "Turkey Red in Blackley: A Chapter in the History of Dyeing."
19. "Journals of the House of Commons, 1786."
20. Chambre de commerce et d'industrie (Rouen), "Observations de la Chambre du commerce de Normandie, sur le traité de commerce entre la France et l'Angleterre (Reprod.)."
21. Cliffe, "Turkey Red in Blackley: A Chapter in the History of Dyeing."
22. Berthoud, *Les indiennes neuchâteloises*, 29, 56–65.
23. Borel, *Les Borel de Bitche, originaires du Val-de Travers en Suisse*.
24. Musson and Robinson, "Science and Technology in the Industrial Revolution," 344–5.
25. Greysmith, "Patterns, Piracy and Protection in the Textile Printing Industry 1787-1850."
26. Nieto-Galan, *Colouring Textiles : A History of Natural Dyestuffs in Industrial Europe*, 128; Walker, "The Life and Influence of Professor J.J. Hummel."
27. Hargreaves, *Messrs. Hargreaves' Calico Print Works at Accrington, and Recollections of Broad Oak*; Jacqué et al., *Andrinople: Le rouge magnifique*, 55.

28. Ashmore, *The Industrial Archaeology of Lancashire*, 296.
29. "Book of Dye Recipes and Samples of Printed Cotton, Made by Foxhill Bank Printworks, England 1830s-1840s."
30. O'Neill, "The Printing and Dyeing of Calico, Silk, and Woollen Fabrics."
31. O'Neill.
32. Gallacher, "The Secret of the Vale."
33. Eyre-Todd, *History of Glasgow*; Biggam, "Rags to Riches"; Quinton, "Turkey Red and the Slave Economy"; Arthur et al., *Seeing Red: Scotland's Exotic Textile Heritage*; Burns, "Parish of Barony of Glasgow"; Tarrant, "The Turkey Red Dyeing Industry in the Vale of Leven"; Tuckett and Nenadic, "Colouring the Nation: A New In-Depth Study of the Turkey Red Pattern Books in the National Museums Scotland"; Gregor, "Parish of Bonhill"; MacKay, *Bleachfields, Printfields and Turkey Red*; Stewart, "Parish of Bonhil"; Peel, "Turkey Red Dyeing in Scotland Its Heyday and Decline"; Johnston, "The Secret of Turkey Red Technology Transfer with a Scottish Connection"; "V. Account of the Process Followed by M. Pierre Jaques Papillon for Dyeing Turkey Red"; Anderson, "Parish of Blantyre"; Brown, "Parish of Rutherglen"; Bremner, *The Industries of Scotland: Their Rise, Progress, and Present Condition*; William Sterling & Sons, "Private Ledger"; Fryer, *Behind the Vale*; Gallacher, "The Secret of the Vale"; Nenadic, "Selling Printed Cottons in Mid-Nineteenth-Century India: John Matheson of Glasgow and Scottish Turkey Red"; Nenadic and Tuckett, *Colouring the Nation: The Turkey Red Printed Cotton Industry in Scotland c. 1840-1940.*
34. Raveux, "Spaces and Technologies in the Cotton Industry in the Seventeenth and Eighteenth Centuries: The Example of Printed Calicoes in Marseilles."
35. Robinson, *A History of Printed Textiles*, 121; Berthoud, *Les indiennes neuchâteloises*, 12–16.
36. Schaefer, "The History of Turkey Red Dyeing"; Oberholzer-Hofmann, *Die Rotfarb Uznach : hundert Jahre im Besitze der Familie Hofmann*; Knoepfli, "Die Sulzersche Rotfarb und Kattun-Druckerei zu Aadorf."
37. Mazzolini-Trümpy, "Von Fabrikanten, Teilhabern, Drogenhändlern und Versicherungsagenten in der Glarner Industrielandschaft des 19. Jahrhunderts," 70; Persoz, *Traité théorique et pratique de l'impression des tissus*.
38. Fairlie, "Dyestuffs in the Eighteenth Century."
39. Berthoud, *Les indiennes neuchâteloises*.
40. Raveux, "Spaces and Technologies in the Cotton Industry in the Seventeenth and Eighteenth Centuries: The Example of Printed Calicoes in Marseilles."
41. Storey, *The Thames and Hudson Manual of Dyes and Fabrics*.
42. Macquer, *Dictionnaire portatif des arts et métiers: contenant en abrégé l'histoire, la description & la police des arts et métiers, des fabriques et manufactures de France et des pays étrangers*.
43. Verbong, "Turksrood."
44. Haller, "The Chemistry and Technique of Turkey Red Dyeing."
45. Verbong, "Turksrood."
46. Arthur et al., *Seeing Red: Scotland's Exotic Textile Heritage*, 16.
47. Lopez, "The Transition from Natural Madder to Synthetic Alizarine in the American Textile Industry, 1870-1890."
48. Ellis, *The Country Dyer's Assistant*.
49. Adrosko and Smith, *Natural Dyes and Home Dyeing*, 21.
50. Bemiss, *The Dyer's Companion*.
51. Smith, "Four Aspects of Turkey Red."

52. Brackman, *Clues in the Calico: A Guide to Identifying and Dating Antique Quilts*, 62.

53. Washington, "On Dyeing Turkey Red."

54. Rauch, *John Rauch's Receipts on Dyeing, in a Series of Letters to a Friend. Containing Correct and Exact Copies of All His Best Receipts on Dyeing Cotton and Woollen Goods; Obtained and Improved by Him, during Twelve Years Practice, at Different Manufactories in Switzerland, France, Germany and America. Also, a true Description of His Invented Substitute for Woad, Being a Cheap and Preferable Material, and the Produce of this Country*; Rauch and Sherman, "John Rauch's Receipts on Dyeing Cotton & Woolen."; Thorsen, "The Merchants and the Dyers : The Rise of a Dyeing Service Industry in Massachusetts and New."

55. Cooper, *A Practical Treatise on Dyeing, and Callicoe Printing: Exhibiting the Processes in the French, German, English, and American Practice of Fixing Colours on Woollen, Cotton, Silk, and Linen.*

56. Musson and Robinson, "Science and Technology in the Industrial Revolution," 347.

57. Storey, "Just New From the Mills - Printed Cottons in Victorian America."

58. Lopez, "The Transition from Natural Madder to Synthetic Alizarine in the American Textile Industry, 1870-1890."

59. Arthur et al., *Seeing Red: Scotland's Exotic Textile Heritage.*

60. Lopez, "The Transition from Natural Madder to Synthetic Alizarine in the American Textile Industry, 1870-1890."

61. Russell, "Personal Conversation."

62. Foster, *Lamb's Textile Industries of the United States: Embracing Biographical Sketches of Prominent Men and a Historical Resume of the Progress of Textile Manufacture from the Earliest Records to the Present Time*, 169.

63. Storey, "Just New From the Mills - Printed Cottons in Victorian America."

64. "S.H. Greene & Sons Records."

65. Arthur et al., *Seeing Red: Scotland's Exotic Textile Heritage*, 18.

66. "Turkey Red Dyeing Company."

67. Higgins, *Dyeing in Germany and America, with Notes on Colour Production.*, 49–51.

68. Ure, *A Dictionary of Arts, Manufactures, and Mines.*

69. Sandberg, *The Red Dyes.*

70. Wadsworth and De Lacy Mann, *The Cotton Trade and Industrial Lancashire, 1600-1780*, 183.

71. O'Neill, "The Printing and Dyeing of Calico, Silk, and Woollen Fabrics."

72. Dépierre, *Traité de la teinture et de l'impression des matières colorantes artificielles. 2me partie : L'Alizarine artificielle et ses dérivés*, 2:391.

73. Schaefer, "The History of Turkey Red Dyeing."

74. "Reports of the United States Commissioners to the Universal Exposition of 1889 at Paris."

75. Factories Inquiry Commission, "Supplementary Report of the Central Board of His Majesty's Commissioners Appointed to Collect Information in the Manufacturing Districts, as to the Employment of Children in Factories, and as to the Propriety and Means of Curtailing the Hours of Their Lab."

76. Anderson, "Parish of Blantyre."

77. Arthur et al., *Seeing Red: Scotland's Exotic Textile Heritage*, 7.

78. Biggam, "Rags to Riches."

79. Le Pileur d'Apligny, *L'art de la teinture des fils et étoffes de coton.*

80. Bremner, *The Industries of Scotland: Their Rise, Progress, and Present Condition.*

81. Lopez, "The Transition from Natural Madder to Synthetic Alizarine in the American Textile Industry, 1870-1890," 113–16.
82. Arthur et al., *Seeing Red: Scotland's Exotic Textile Heritage*.
83. Tuckett and Nenadic, "Colouring the Nation: A New In-Depth Study of the Turkey Red Pattern Books in the National Museums Scotland."
84. İnalcik, "When and How British Cotton Goods Invaded the Levant Markets."
85. St Clair, *The Golden Thread: How Fabric Changed History*, 168.
86. Robinson, *A History of Printed Textiles*.
87. St Clair, *The Golden Thread: How Fabric Changed History*, 168–71.
88. "On Bandana Handkerchiefs."
89. Mukund, "Indian Textile Industry in 17th and 18th Centuries: Structure, Organisation and Responses."
90. Nenadic, "Selling Printed Cottons in Mid-Nineteenth-Century India: John Matheson of Glasgow and Scottish Turkey Red."
91. Bassett, *The Technological Indian*, 37–9.

Chapter 6

1. Nenadic and Tuckett, *Colouring the Nation: The Turkey Red Printed Cotton Industry in Scotland c. 1840-1940*, 24.
2. Tuckett and Nenadic, "Colouring the Nation: A New In-Depth Study of the Turkey Red Pattern Books in the National Museums Scotland."
3. Wilder, *Little House in the Big Woods*.
4. Frank, "Publisher's Notices," 66.
5. Le département Hauts-de-seine, "Archives de la Planète."
6. Wadsworth and De Lacy Mann, *The Cotton Trade and Industrial Lancashire, 1600-1780*.
7. "Dublin Exchange."
8. "Alex Ramsay and Co."
9. Lemire, "'A Good Stock of Cloaths': The Changing Market for Cotton Clothing in Britain, 1750-1800."
10. Arthur et al., *Seeing Red: Scotland's Exotic Textile Heritage*, 16–18.
11. "John Jordan Jr . . . Merchandize."
12. "Blue & Black Dying."
13. "First Fall Goods."
14. "New Goods—Andrews and Brothers."
15. "New Spring & Summer Goods at the Cheap Cash Store."
16. "New Goods—J.W. Armstrong & Co."
17. "Dry Goods."
18. "New Store! New Goods! - W.P. White & Co."
19. "Local Notices."
20. "A Colossal Mart - Osborne, Kespohl & Co."
21. "Some Bargains That Will Fetch Buyers in a Hurry."
22. "For Sale at Dring's Adobie Store."

23. "Rice & Co."
24. "For Sale . . . The Cargo of the 'Humphrey Nelson.'"
25. "Fresh Goods."
26. "T.G. Pitman."
27. "Quotation of Cotton Piece Goods, Nippon Kokusan Co., Ltd."
28. O'Connor, "Four Aspects of Turkey Red: The Process and Early History."
29. Montgomery Ward and Company, "Catalogue No. 13 Spring and Summer 1875," 14–16.
30. Tuckett and Nenadic, "Colouring the Nation: A New In-Depth Study of the Turkey Red Pattern Books in the National Museums Scotland"; Smith, "Four Aspects of Turkey Red."
31. Bremner, *The Industries of Scotland: Their Rise, Progress, and Present Condition.*
32. Tuckett and Nenadic, "Colouring the Nation: A New In-Depth Study of the Turkey Red Pattern Books in the National Museums Scotland"; Eastop, "New Ways of Engaging with Historic Textiles: Interactive Images Online."
33. Arthur et al., *Seeing Red: Scotland's Exotic Textile Heritage*, 19–20.
34. Wood, "Turkey Red."
35. Peel, "Turkey Red Dyeing in Scotland Its Heyday and Decline."
36. Smith, "Four Aspects of Turkey Red."
37. Brackman, *Clues in the Calico: A Guide to Identifying and Dating Antique Quilts*, 105–6.
38. Tuckett and Nenadic, "Colouring the Nation: A New In-Depth Study of the Turkey Red Pattern Books in the National Museums Scotland."
39. Brackman, *Clues in the Calico: A Guide to Identifying and Dating Antique Quilts*; Welters and Ordoñez, *Down by the Old Mill Stream: Quilts in Rhode Island.*
40. Brackman, *Clues in the Calico: A Guide to Identifying and Dating Antique Quilts*; Arthur et al., *Seeing Red: Scotland's Exotic Textile Heritage.*
41. Biscula, Schattenburg-Raymond, and du Preez, "Hawaiian Barkcloth from the Bishop Museum Collections: A Characterization of Materials and Techniques in Collaboration with Modern Practitioners to Effect Preservation of a Traditional Cultural Practice"; Arthur, "Cultural Authentication of Hawaiian Quilting in the Early 19th Century"; Kent, *Treasury of Hawaiian Words in One Hundred and One Categories*, 151.
42. Arthur et al., *Seeing Red: Scotland's Exotic Textile Heritage*, 6.
43. Quinton, "Turkey Red and the Slave Economy."
44. Cole-Leonard, "Maya Angelou's 'Hallelujah! The Welcome Table, A Lifetime of Memories with Recipes' as Evocative Text, or, 'Ain't' Jemima's Recipes."
45. Adams, "Four Aspects of Turkey Red: Turkey Red in Quilts and Clothing."
46. Smith, "Four Aspects of Turkey Red."
47. Peel, "Turkey Red Dyeing in Scotland Its Heyday and Decline."
48. Smith, "Four Aspects of Turkey Red."
49. Brackman, "Turkey Red"; Mauro and Fiorenzani, "Sul 'Follone di Garibaldi' esistente a Prato."

References

"A Colossal Mart—Osborne, Kespohl & Co." *Lincoln Daily Evening News*. September 29, 1883.

"Alex Ramsay and Co." *The Dublin Journal*. April 2, 1799.

"Blue & Black Dying." *The Maryland Herald, and Hager's-Town Weekly Advertiser*. May 9, 1806.

"Book of Dye Recipes and Samples of Printed Cotton, Made by Foxhill Bank Printworks, England 1830s–1840s." London, n.d., T.8-1978. Victoria & Albert Museum Collection.

"Correspondence with Marta Turok," 2021.

"Discharging of Turkey-Red." *The Glasgow Mechanics' Magazine* and Annals of Philosophy 1, no. 11 (1824): 176.

"Dry Goods." *The Boston Morning Post*. February 20, 1840.

"Dublin Exchange." *The Dublin Journal*. November 21, 1778.

"Dyeing Turkey Red." *Scientific American* 24, no. 4 (1871): 57.

"First Fall Goods." *Savannah Republican*. December 24, 1816.

"First Friday + Barbara Brackman's Lecture—The Quilt Detective: Clues in Turkey Red." International Quilt Museum, 2022. Accessed March 6, 2023. https://www.internationalquiltmuseum.org/first-friday-barbara-brackmans-lecture-quilt-detective-clues-turkey-red

"For Sale at Dring's Adobie Store." *The Californian*. January 11, 1849.

"For Sale . . . The Cargo of the 'Humphrey Nelson.'" *Honolulu Pacific Commercial Advertiser*. January 5, 1860.

"Fresh Goods." *Hobart Town Gazette and Van Diemen's Land Advertiser*. November 30, 1822.

"John Jordan Jr . . . Merchandize." *Kentucky Gazette and General Advertiser*. January 18, 1803.

"Journals of the House of Commons, 1786." Vol. 41. London, 1803.

"Local Notices." *Columbus Journal*. February 26, 1879.

"Mémoire contenant le procédé de la teinture du coton rouge-incarnat d'Andrinople sur le coton filé." Paris: Imprimerie Royale, 1765.

"New Goods—Andrews and Brothers." *Tuscaloosa Alabama State Intelligencer*. February 16, 1831.

"New Goods—J.W. Armstrong & Co." *Kingston Upper Canada Herald*. June 19, 1833.

"New Spring & Summer Goods at the Cheap Cash Store." *Tarboro North Carolina Free Press*. April 27, 1833.

"New Store! New Goods!—W.P. White & Co." *The Rising Sun Indiana Whig*. January 19, 1850.

"Notices to Correspondents: Bandana Controversy." *The Glasgow Mechanics' Magazine* 1, no. 14 (1824): 224.

"Notices to Correspondents. War on Turkey-Red." *The Glasgow Mechanics' Magazine and Annals of Philosophy* 1, no. 10 (1824): 160.

"Notices to Correspondents." *The Glasgow Mechanics' Magazine and Annals of Philosophy* 1, no. 13 (1824): 208.

"On Bandana Handkerchiefs." *The Saturday Magazine* 16, no. 493 (1840): 91–3.

"On the Process for Discharging Turkey-Red." *The Glasgow Mechanics' Magazine* 1, no. 8 (1824): 117–18.

"Quotation of Cotton Piece Goods, Nippon Kokusan Co., Ltd." *The Japan Advertiser*. November 26, 1920.

"Reports of the United States Commissioners to the Universal Exposition of 1889 at Paris." *The Executive Documents of the House of Representatives for the First Session of the Fifty-First Congress 1889–90*. Vol. 39. Washington DC, 1893.

"Rice & Co." *The Polynesian*. December 29, 1855.

"S.H. Greene & Sons Records." Rhode Island Historical Society. Accessed May 9, 2021. https://www.rihs.org/mssinv/Mss953.htm

"Some Bargains That Will Fetch Buyers in a Hurry." *Goshen Weekly News*. March 23, 1895.

"Specification of the Patent Granted to James Thomson, of Primrose Hill, near Clithero, in the County of Lancaster, Calico Printer; for a Method of Producing Patterns on Cloth Previously Dyed Turkey Red, and Made of Cotton or Linen, or Both." *The Repertory of Arts, Manufactures, and Agriculture* 23, no. 136 (1813): 193–200.

"Sunday School Object Lesson." *The Christian Recorder*. Philadelphia, PA, August 15, 1868.

"T.G. Pitman." *The Sydney Monitor*. May 26, 1826.

"The Manufacture of Articles of Commerce from Blood." In *Twelfth Report of the Medical Officer of the Privy Council (1869)*, 211–23. London: HM Stationery Office, 1874.

"The Red Society of Glasgow: Minute Book." Glasgow, n.d., T-TH15-26. The Mitchell Library Collection.

"The Red Society of Glasgow: Seal of Cause from the Magistrates and Town Council of Glasgow to James Bogle, Preces, David Fairie, Collector, James Campbell, Alexander Scott, Gershom Robertson, James Marshall, William Killoch, William McFarland, William Fin." Glasgow, 1759, T-TH15-25. The Mitchell Library Collection

"Turkey Red Dyeing Company." *The Boston Globe*. November 3, 1886.

"V. Account of the Process Followed by M. Pierre Jaques Papillon for Dyeing Turkey Red." *Philosophical Magazine Series 1* 18, no. 69 (1804): 43–7.

Achaya, K. T. "Chemical Derivatives of Castor Oil." *Journal of the American Oil Chemists' Society* 48, no. 11 (1971): 758–63.

Adams, Nettie. "Clothing and Textiles of Ottoman Egypt: Examples from Art and Archaeology." In *Textile Society of America Symposium Proceedings*, 81–90, 1992.

Adams, Pauline. "Four Aspects of Turkey Red: Turkey Red in Quilts and Clothing." *Quilt Studies* 1 (1999).

Adrosko, Rita J, and Margaret Smith. *Natural Dyes and Home Dyeing*. New York, NY: Dover Publications, 1971.

Ahmad, Ishtiaque, and Jasbhinder Singh. "Surface Active Properties of Sulfonated Isoricinoleic Acid." *Journal of the American Oil Chemists' Society* 67, no. 4 (1990): 205–8.

Aikin, John. *A Description of the Country from Thirty to Forty Miles Round Manchester*. London: John Stockdale, 1795.

Anderson, Ernest, and Harry J Lowe. "The Composition of Flaxseed Mucilage." *The Journal of Biological Chemistry* 168, no. 1 (1947): 289–97.

Anderson, James. "Parish of Blantyre." In *The New Statistical Account of Scotland*. Edinburgh: William Blackwood and Sons, 1845.

Andriessen, Robert. "Turkish Red & More." YouTube: TextielMuseum, 2013. Accessed March 6, 2023. https://www.youtube.com/watch?v=0MaSUFhca7Q

Archibald Orr Ewing & Co. "Turkey Red Dyeing Calculation Book AOE Lennoxbank." Glasgow, n.d. University of Glasgow Archives and Special Collections. GB248 UGD 13/4/1.

Arthur, Linda. "Cultural Authentication of Hawaiian Quilting in the Early 19th Century." *Clothing and Textiles Research Journal* 29, no. 2 (2011): 103–18.

Arthur, Liz, John R Hume, Mary Schoeser, and Lindsay Taylor. *Seeing Red: Scotland's Exotic Textile Heritage*. Glasgow: Friends of Glasgow Museums, 2007.

Ashmore, Owen. *The Industrial Archaeology of Lancashire*. Newton Abbot: David & Charles, 1969.

Ashmore, Sonia, Steven Cohen, Rosemary Crill, Avalon Fotheringham, Barbara Karl, Pramod Kumar KG, Divia Patel, Anamika Pathak, and Aurelie Samuel. *The Fabric of India*. Edited by Rosemary Crill. London: V&A Publishing, 2015.

Atayolu, Sabri. "Beitrage zur Geschichte der Farberei aus turkischen Archiven." *Melliand Textilberichte* 17 (1936): 74–5.

Auerbach, G. *Anthracen*. Translated by William Crookes. London: Longmans, Green, and Co., 1877.

Baker, George Percival. "East Indian Hand-Painted Calicoes of the Seventeenth and Eighteenth Centuries, and Their Influence on the Tinctorial Arts of Europe." *Journal of the Royal Society of Arts* 64, no. 3313 (May 19, 1916): 482–92.

Bamford, Debbie. "Turkey Red." *Journal for Weavers, Spinners and Dyers*, no. Winter (2016): 25–7.

Barry Barnett, E de. *Anthracene and Anthraquinone*. London: Ballière, Tindall and Cox, 1921.

Bassett, Ross. *The Technological Indian*. Cambridge, MA and London, England: Harvard University Press, 2016.

Beaujard, Philippe. "Gujarat and Long-Distance Trade in the Indian Ocean Region before the Sixteenth Century." In *Transregional Trade and Traders: Situating Gujarat in the Indian Ocean from Early Times to 1900*, edited by E. A. Alpers and C. Goswami. Delhi: Oxford University Press, 2019.

Beke, X. *Notes and Queries: A Medium of Intercommunication for Literary Men, General Readers, Etc.* Vol. 7. London: John C. Francis, 1890.

Bemiss, Elijah. *The Dyer's Companion*. New London, CT: Printed by Cady & Eells for the author, 1806.

Bergerhoff, G, and Christian-Heinrich Wunderlich. "Crystal Structure of di-μ-oxo-bis(bis(1,2,4-trihydroxyanthrachinonato)aluminium-(aquabis(dimethylformamide)calcium)), $Al_2O_4(C_{14}H_6O_5)_4Ca_2$ $(HCON(CH_3)_2)_4(H_2O)_2$, (purpurine complex)." *Zeitschrift Für Kristallographie—Crystalline Materials* 207, no. 1 (1993): 189–92.

Berthollet, Claude-Louis. *Elements of the Art of Dyeing*. Translated by William Hamilton. London: Stephen Couchman, 1791.

Berthoud, Dorette. *Les indiennes neuchâteloises*. Boudry: A la Baconnière, 1951.

Bien, Hans-Samuel, Josef Stawitz, and Klaus Wunderlich. "Anthraquinone Dyes and Intermediates." In *Ullmann's Encyclopedia of Industrial Chemistry*, 513–78. John Wiley and Sons, Inc., 2012.

Biggam, Carole. "Rags to Riches." Helensburgh, 2008.

Biscula, Christina, Lisa Schattenburg-Raymond, and Kamalu du Preez. "Hawaiian Barkcloth from the Bishop Museum Collections: A Characterization of Materials and Techniques in Collaboration with Modern Practitioners to Effect Preservation of a Traditional Cultural Practice." *Materials Research Society Symposium Proceedings* 1656 (2014).

Blackburn, Richard S. "Natural Dyes in Madder (Rubia Spp.) and Their Extraction and Analysis in Historical Textiles." *Coloration Technology* 133, no. 6 (2017): 449–62.

Boot, Caroline. "Turkish Red & More : Formafantasma." TextielMuseum, 2013. Accessed March 6, 2023. https://textielmuseum.nl/turkish-red-more-formafantasma/

Borel, Hermann Henry. *Les Borel de Bitche, originaires du Val-de Travers en Swisse*. Genève: A. Kundig, 1917.

Bossert, Roy G. "The Metallic Soaps." *Journal of Chemical Education* 27, no. 1 (January 1, 1950): 10.

Brackman, Barbara. "Turkey Red Tour." Barbara Brackman's Material Culture, 2016. Accessed March 6, 2023. http://barbarabrackman.blogspot.com/2016/10/turkey-red-tour.html

Brackman, Barbara. "Turkey Red." Barbara Brackman's Material Culture, 2010. Accessed March 6, 2023. http://barbarabrackman.blogspot.com/2010/09/turkey-red.html

Brackman, Barbara. *Clues in the Calico: A Guide to Identifying and Dating Antique Quilts.* Charlottesville, VA: Howell Press, 1989.

Brandt, Charles, Joseph Dépierre, Albert Scheurer, Henri Schmid, Léon Stamm, Félix Weber, and Alphonse Wehrlin. "Sulfoléates et rouge turc par le procédé rapide, priorités." *Bulletin de la Société Industrielle de Mulhouse* 79, no. September (1909).

Brandt, Charles. "Rapport présenté au nom du comité de chimie par M. Brandt, sur la valeur comparée de l'alizarine artificielle et de la garance." *Bulletin de la Société Industrielle de Mulhouse* 43 (1873).

Bremner, David. *The Industries of Scotland: Their Rise, Progress, and Present Condition.* Edinburgh: Adam and Charles Black, 1869.

Brennan, Louis N, John Tannahill, and James P Percy. "Turkey Red Process for Cotton Yarn." Paisley: Paisley Central Library, 1943.

Brown, Peter. "Parish of Rutherglen." In *The New Statistical Account of Scotland*, 386–7. Edinburgh: William Blackwood and Sons, 1845.

Burnett, A.R., and R.H. Thomson. "Naturally Occurring Quinones. Part XV. Biogenesis of the Anthraquinones in Rubia Tinctorum L. (Madder)." *Journal of the Chemical Society C: Organic*, 1968, 2437–41.

Burns, John. "Parish of Barony of Glasgow." In *The Statistical Account of Scotland*, edited by John Sinclair, 109–26. Edinburgh: William Creech, 1794.

Burton, Donald, and George F Robertshaw. *Sulphated Oils and Allied Products.* New York, NY: Chemical Publishing Co., Inc., 1940.

Cain, John Cannell, and Jocelyn Field Thorpe. *The Synthetic Dyestuffs and the Intermediate Products from Which They Are Derived.* London: Charles Griffin & Company, Limited, 1905.

Campbell, David. "Mr. Campbell's Proof of Claims." *The Glasgow Mechanics' Magazine* 2, no. 48 (1824): 287–88.

Campbell, David. "Statement of Claims to the Invention of the Process for Discharging Turkey-Red." *The Glasgow Mechanics' Magazine* 3, no. 89 (1825): 45–7.

Cardon, Dominique. *Natural Dyes: Sources, Tradition, Technology and Science.* Translated by Catherine Higgitt. London: Archetype Publications Ltd., 2007.

Caro, Heinrich, Charles Graebe, and Charles Liebermann. Preparing Coloring Matters. GB18691936. United Kingdom: Office of the Commissioners for Patents, issued 1870.

Carruthers, W.S. "A New Process of Dyeing Turkey Red." *Journal of the Society of Dyers and Colourists*, 1911, 123–6.

Chambre de commerce et d'industrie (Rouen). "Observations de la Chambre du commerce de Normandie, sur le traité de commerce entre la France et l'Angleterre (Reprod.)." Rouen, 1788.

Chao, Ju-Kua. *Chau Ju-Kua: His Work on the Chinese and Arab Trade in the Twelfth and Thirteenth Centuries, Entitled Chu-Fan-Chï.* Translated by Friedrich Hirth and W.W. Rockhill. St. Petersburg: Imperial Academy of Sciences, 1911.

Chaptal, Jean-Antoine. *L'art de la teinture du coton en rouge.* Paris: Deterville, 1807.

Chateau, Theodore. "Critical and Historical Notes Concerning the Production of Adrianople or Turkey Red, and the Theory of This Colour No. 3." Edited by Charles O'Neill. *The Textile Colourist* 1, no. March (1876): 172–78.

Chateau, Theodore. "Critical and Historical Notes Concerning the Production of Adrianople or Turkey Red, and the Theory of This Colour No. 4." Edited by Charles O'Neill. *The Textile Colourist* 1, no. April (1876): 217–31.

Chateau, Theodore. "Critical and Historical Notes Concerning the Production of Adrianople or Turkey Red, and the Theory of This Colour No. 6." Edited by Charles O'Neill. *The Textile Colourist* 1, no. June (1876): 384–97.

Chateau, Theodore. "Critical and Historical Notes Concerning the Production of Adrianople or Turkey Red, and the Theory of This Colour No. 7." Edited by Charles O'Neill. *The Textile Colourist* 7, no. July (1876): 27–33.

Chateau, Theodore. "Critical and Historical Notes Concerning the Production of Adrianople or Turkey Red, and the Theory of This Colour No. 9." Edited by Charles O'Neill. *The Textile Colourist* 2, no. September (1876): 131–41.

Chateau, Theodore. "Critical and Historical Notes Concerning the Production of Adrianople or Turkey Red, and the Theory of This Colour No. 10." Edited by Charles O'Neill. *The Textile Colourist* 2, no. October (1876): 191–200.

Chateau, Theodore. "Critical and Historical Notes Concerning the Production of Adrianople or Turkey Red, and the Theory of This Colour No. 11." Edited by Charles O'Neill. *The Textile Colourist* 2, no. November (1876): 262–72.

Chenciner, Robert. *Madder Red: A History of Luxury and Trade.* Surrey: Curzon Press, 2000.

Chettra, Satinderjeet Kaur. "Microscopy and Surface Chemical Investigations of Dyed Cellulose Textiles." University of Nottingham, 2006.

Christie, John. "XII. Notes of Experiments on Artificial Alizarine." In *Proceedings of the Philosophical Society of Glasgow*, Vol. 7. Glasgow: John Smith and Son, 1871.

Clark, Hazel. "The Design and Designing of Lancashire Printed Calicoes during the First Half of the 19th Century." *Textile History* 15, no. 1 (1984): 101–18.

Clements, W., and J. Sadler. *The New Handmaid to Arts Sciences Agriculture.* London, 1790.

Clibbens, D.A. "Some Research Problems in Cotton Bleaching and Dyeing." *Journal of the Society of Dyers and Colourists* 41, no. 7 (1925): 249–50.

Cliffe, W.H. "Turkey Red in Blackley: A Chapter in the History of Dyeing." *Manchester Literary and Philosophical Society Memoirs and Proceedings* 118 (1975): 80–92.

Clow, Archibald., and Nan L. Clow. "The Chemical Revolution: A Contribution to Social Technology." London: Batchworth Press, 1952.

Cole-Leonard, Natasha. "Maya Angelou's 'Hallelujah! The Welcome Table, A Lifetime of Memories with Recipes' as Evocative Text, or, 'Ain't' Jemima's Recipes." *The Langston Hughes Review* 19, no. Spring (2005): 66–69.

Colin, J-J, and P-J Robiquet. "Nouvelles recherches sur la matière colorante de la garance." *Annales de Chimie et de Physique* 34 (1827): 225–53.

Collin, Dr. "Turkey Red Process for Cotton Yarn." Paisley: Paisley Central Library, n.d.

Conseil d'Etat France. "Arrêt du conseil d'état qui, en autorisant la manufacture de draps de soie, laine, ratines et peluches, établie à Montmartre par le Sieur Quinquet, lui permet de faire teindre dans ladite manufacture, en grand et bon teint et en rouge d'Andrinople, toutes." Paris: Imprimerie Royale, 1776.

Cooksey, Christopher J, and Alan T Dronsfield. "Edward Schunck: Forgotten Dyestuffs Chemist?" In *Dyes in History and Archaeology* 21, edited by Jo Kirby, Christopher J Cooksey, Maarten R. van Bommel, Anita Quye, and Richard Atkinson, 189–207. London: Archetype Publications Ltd., 2008.

Cooper, Thomas. *A Practical Treatise on Dyeing, and Callicoe Printing: Exhibiting the Processes in the French, German, English, and American Practice of Fixing Colours on Woollen, Cotton, Silk, and Linen*. Philadelphia, PA: Thomas Dobson, 1815.

Crookes, William. *A Practical Handbook of Dyeing and Calico-Printing*. London: Longmans, Green and Co., 1874.

Crookes, William. *Dyeing and Tissue Printing*. Edited by H Trueman Wood. London: George Bell and Sons, 1882.

Cunningham, Anthony B., I. Made Maduarta, Jean Howe, W. Ingram, and Steven Jansen. "Hanging by a Thread: Natural Metallic Mordant Processes in Traditional Indonesian Textiles." *Economic Botany* 65, no. 3 (2011): 241–59.

Cuoco, Guillaume, Carole Mathe, Paul Archier, and Cathy Vieillescazes. "Characterization of Madder and Garancine in Historic French Red Materials by Liquid Chromatography-Photodiode Array Detection." *Journal of Cultural Heritage* 12, no. 1 (2011): 98–104.

D.B. "On Dyeing Red." *The Glasgow Mechanics' Magazine* 1, no. 5 (1824): 76.

Decaisne, Joseph. "XXIX.—On the Root of the Madder." *Annals and Magazine of Natural History* 1, no. 4 (June 1838): 267–74.

Decaisne, Joseph. *Recherches anatomiques et physiologiques sur la garance. Annals and Magazine of Natural History*. Brussels: M. Hayez, 1837.

Delamare, François, Bernard Monasse, and Michel Garcia. "The Role of Aluminium as a Mordant for Cellulose Dyeing with Alizarin: A Numerical Approach." In *The Diversity of Dyes in History and Archaeology*, edited by Jo Kirby, Dominique Cardon, Chris Cooksey, Vanessa Habib, Richard Laursen, Anita Quye, Terry Schaeffer, André Verhecken, and Maarten R. van Bommel. London: Archetype Publications Ltd., 2017.

Dépierre, Joseph. *Traité de la teinture et de l'impression des matières colorantes artificielles. 2me partie : L'Alizarine artificielle et ses dérivés*. Vol. 2. Paris: Baudry et Cie, 1892.

Derksen, Goverdina C. H., Gerrit P. Lelyveld, Teris A. van Beek, Anthony Capelle, and Æde de Groot. "Two Validated HPLC Methods for the Quantification of Alizarin and Other Anthraquinones in Rubia Tinctorum Cultivars." *Phytochemical Analysis* 15, no. 6 (2004): 397–406.

Derksen, Goverdina C. H., Martijn Naayer, Teris A. van Beek, Anthony Capelle, Ingrid K. Haaksman, Henk A. van Doren, and Æde de Groot. "Chemical and Enzymatic Hydrolysis of Anthraquinone Glycosides from Madder Roots." *Phytochemical Analysis* 14, no. 3 (2003): 137–44.

Driessen, Félix. "Étude sur le rouge turc, ancien procédé." *Bulletin de la Société Industrielle de Mulhouse* 72 (1902): 163–80.

Dronsfield, Alan T, Trevor Brown, and Christopher J Cooksey. "Synthetic Alizarin- the Dye That Changed History." In *Dyes in History and Archaeology* 16/17. London: Archetype Publications, 2001.

Dumas, Jean-Baptiste. *Traité de chimie, appliquée aux arts*. Vol. 8. Traité de chimie, appliquée aux arts. Paris: Béchet Jeune, 1846.

Dunbar, C. "Modern Developments in Textile Chemicals for Dyeing and Finishing." *Journal of the Society of Dyers and Colourists* 50, no. 10 (1934): 309–16.

Eastop, Dinah. "New Ways of Engaging with Historic Textiles: Interactive Images Online." *Textile History* 47, no. 1 (2016): 82–93.

Ellis, Asa. *The Country Dyer's Assistant*. Brookfield, MA, 1798.

Espinosa-Jiménez, M, F González-Caballero, C F González-Fernández, and G Pardo. "The Adsorption of Tannic Acid on Hydrophilic Cotton and Its Effect on the Electrokinetic Properties of This Cellulose Fibre in a Cationic Dye Solution." *Acta Polymerica* 38, no. 1 (1987): 96–100.

Eyre-Todd, George. *History of Glasgow*. Vol. 3. Glasgow: Jackson, Wylie & Co., 1934.

Factories Inquiry Commission. "Supplementary Report of the Central Board of His Majesty's Commissioners Appointed to Collect Information in the Manufacturing Districts, as to the Employment of Children in Factories, and as to the Propriety and Means of Curtailing the Hours of Their Lab." London, 1834.

Fairlie, Susan. "Dyestuffs in the Eighteenth Century." *The Economic History Review* 17, no. 3 (April 1, 1965): 488–510.

Faroqhi, Suraiya. "Declines and Revivals in Textile Production." In *The Cambridge History of Turkey: Volume 3: The Later Ottoman Empire, 1603–1839*, edited by Suraiya N Faroqhi, 3:356–75. Cambridge History of Turkey. Cambridge: Cambridge University Press, 2006.

Faroqhi, Suraiya. "Ottoman Cotton Textiles, 1500s to 1800: The Story of a Success That Did Not Last." In *The Spinning World: A Global History of Cotton Textiles, 1200–1850*, edited by Giorgio Riello and Prasannan Parthasarathi, 89–104. Oxford: Oxford University Press, 2009.

Fereday, Gwen. *Natural Dyes*. Edited by Suzannah Dick, Caroline Brooke Johnson, and Coralie Hepburn. London: The British Museum Press, 2003.

Fierz-David, Hans Eduard, and Max Rutishauser. "Composition and Constitution of Turkey Red." *Journal of the Society of Dyers and Colourists*, 1941, 57–8.

Fieser, Louis F. "The Discovery of Synthetic Alizarin." *Journal of Chemical Education* 7, no. 11 (1930): 2609–33.

Flachat, Jean-Claude. *Observations sur le commerce et sur les arts d'une partie de l'Europe, de l'Asie, de l'Afrique, et même des Indes Orientales*. Vol. 2. Lyon: Jacquenod Père & Rusand, 1766.

Foster, E E. *Lamb's Textile Industries of the United States: Embracing Biographical Sketches of Prominent Men and a Historical Resume of the Progress of Textile Manufacture from the Earliest Records to the Present Time*. James H. Lamb, 1916.

Fotheringham, Avalon. *The Indian Textile Sourcebook*. London: Victoria and Albert Museum/Thames & Hudson, 2019.

Fox, Robert. "Presidential Address: Science, Industry, and the Social Order in Mulhouse, 1798–1871." *The British Journal for the History of Science* 17, no. 2 (1984): 127–68.

Frank, M. "Publisher's Notices." *The Textile Colorist* 2, no. 15 (1880).

Fryer, Linda, ed. *Behind the Vale*. Strathclyde Regional Council, 1995.

Fukasawa, Katsumi. *Toile et commerce du Levant d'Alep à Marseille*. Paris: Centre National de la Recherche Scientifique, 1987.

Gallacher, Tom. "The Secret of the Vale." *The Scots Magazine* March (1993): 302.

Gekas, Athanasios. "A Global History of Ottoman Cotton Textiles, 1600-1850." *EUI Working Papers Max Weber Programme*. San Domenico di Fiesole, 2007.

Gillard, R D, S M Hardman, R G Thomas, and D E Watkinson. "The Mineralization of Fibres in Burial Environments." *Studies in Conservation* 39, no. 2 (March 25, 1994): 132–40.

Girard, P.S. *Mémoire sur l'agriculture, l'industrie et le commerce de l'Égypte*. Paris: Imprimerie Royale, 1822.

Gordon, Paul Francis, and Peter Gregory. *Organic Chemistry in Colour*. Berlin: Springer-Verlag, 1983.

Graebe, C., and C. Liebermann. "Ueber Alizarin und Anthracen." *Berichte der Deutschen Chemischen Gesellschaft* 1, no. 1 (1868): 49–51.

Greene, Susan W. *Wearable Prints, 1760–1860: History, Materials, and Mechanics*. Kent State University Press, 2014.

Gregor, William. "Parish of Bonhill." In *The New Statistical Account of Scotland*, 222–8. Edinburgh: William Blackwood and Sons, 1845.

Greysmith, David. "Patterns, Piracy and Protection in the Textile Printing Industry 1787–1850." *Textile History* 14, no. 2 (1983): 165–94.

Gunstone, Frank D., and F.B. Padley. *Lipid Technologies and Applications*. CRC Press, 1997.

Guy, John S. "One Thing Leads to Another: Indian Textiles and the Early Globalization of Style." In *The Interwoven Globe International Textile Trade, 1500–1800*, edited by Amelia Peck. New York, NY: The Metropolitan Museum of Art, 2013.

H. "On the Process for Discharging Turkey-Red." *The Glasgow Mechanics' Magazine*, no. 9 (1824).

Haller, R. "The Chemistry and Technique of Turkey Red Dyeing." *Ciba Review* 39, no. May (1941): 1417–21.

Hanna, Nelly. *Ottoman Egypt and the Emergence of the Modern World: 1500–1800*. Cairo: Cairo Scholarship Online, 2014.

Hargreaves, B. *Messrs. Hargreaves' Calico Print Works at Accrington, and Recollections of Broad Oak*. Accrington: E. Bowker, 1882.

Harvey, Alexander. "Statement of Claims on the Invention of the Process for Discharging Turkey-Red." *The Glasgow Mechanics' Magazine* 1, no. 13 (1824).

Haussmann, J M. "XLIII. Observations on Maddering; Together with a Simple and Certain Process for Obtaining, with Great Beauty and Fixity, That Colour Known under the Name of the Turkey or Adrianople Red." *Philosophical Magazine Series 1* 12, no. 47 (1802): 260–6.

Haussmann, J M. "XXXII. Observations on Maddering; Together with a Simple and Certain Process for Obtaining, with Great Beauty and Fixity, That Colour Known under the Name of the Turkey or Adrianople Red." *Philosophical Magazine Series 1* 12, no. 47 (1802): 170–5.

Hellot, M, M Macquer, and M Le Pileur d'Apligny. *The Art of Dyeing Wool, Silk, and Cotton*. Translated by Anonymous. London: R. Baldwin, 1789.

Henry, Thomas. "Considerations Relative to the Nature of Wool, Silk, and Cotton, as Objects of the Art of Dying; on the Various Preparations, and Mordants, Requisite for These Different Substances; and on the Nature and Properties of Colouring Matter. Together with Some." In *Memoirs of the Literary and Philosophical Society of Manchester*, 3:343–408. London: Warrington, 1790.

Heyne, Benjamin. *Tracts, Historical and Statistical, on India: With Journals of Several Tours Through Various Parts of the Peninsula : Also, an Account of Sumatra, in a Series of Letters*. Nineteenth Century Collections Online: Mapping the World: Maps and Travel Literature. London: R. Baldwin and Black, Parry and Company, 1814.

Higgins, S H. *Dyeing in Germany and America, with Notes on Colour Production*. 2nd edn. Manchester: The University Press, 1916.

Hill, Robert, and Derek Richter. "Anthraquinone Colouring Matters: Galiosin; Rubiadin Primeveroside," no. 1714 (1936): 1714–19.

Hinckley, C.T. "Everyday Actualities. No. II." *Godey's Lady's Book*. Philadelphia, PA, 1852.

Hobson, Debra K, and David S Wales. "'Green' Dyes." *Biotechnology*, no. February (1998): 42–44.

Hofenk de Graaff, Judith H., Wilma G. Th. Roelofs, and Maarten R. van Bommel. *The Colourful Past*. London and Riggisberg: Archetype Publications and Abegg-Stiftung, 2004.

Holme, Ian. "Sir William Henry Perkin: A Review of His Life, Work and Legacy." *Coloration Technology* 122, no. 5 (2006): 235–51.

Hornix, Willem J. "From Process to Plant: Innovation in the Early Artificial Dye Industry." *The British Journal for the History of Science* 25, no. 1 (1992): 65–90.

Huguenot Society of London. *Proceedings of the Huguenot Society of London*. Vol. 1. Huguenot Society of London, 1885.

Hummel, John James. *The Dyeing of Textile Fabrics*. London: Cassell and Company Ltd., 1885.

Hurst, George H. "Recent Progress in Dyeing." *Journal of the Society of Dyers and Colourists*, no. April (1894): 70–4.

Hurst, George H. *Textile Soaps and Oils*. London: Scott, Greenwood & Son, 1904.

İnalcik, Halil. "When and How British Cotton Goods Invaded the Levant Markets." In *The Ottoman Empire and the World-Economy*, edited by Huri İslamoğlu-İnan, 374–83. Cambridge: Cambridge University Press, 1987.

Jacqué, Jacqueline, Jean-François Keller, Joyce Storey, Naomi E A Tarrant, Daniel Fues, Bruno Sueur, Jean- Jacques Zundel, Maurice Binder, and Nadine Botella. *Andrinople: Le rouge magnifique*. Edited by Jacqueline Jacqué. Paris: Éditions de La Martinière, 1995.

Jansen, Steven, Toshihiro Watanabe, and Erik Smets. "Aluminium Accumulation in Leaves of 127 Species in Melastomataceae, with Comments on the Order Myrtales." *Annals of Botany* 90, no. 1 (July 1, 2002): 53–64.

Jenny, M. "Mémoire sur la fabrication du rouge d'Andrinople, présenté par M. Jenny, et traduit de l'allemand par M. Rosenstiehl." *Bulletin de la Société Industrielle de Mulhouse* 38 (1868): 747–837.

Johnston, W. T. "The Secret of Turkey Red Technology Transfer with a Scottish Connection." *Biotechnic and Histochemistry* 85, no. 5 (2010): 295–303.

Karadag, Recep, and Emre Dolen. "Re-Examination of Turkey Red." *Annali Di Chimica* 97 (2007): 583–9.

Katsiardi-Hering, Olga. "The Allure of Red Cotton Yarn, and How It Came to Vienna: Associations of Greek Artisans and Merchants Operating between the Ottoman and Hapsburg Empires." In *Merchants in the Ottoman Empire*, edited by Suraiya Faroqhi and Gilles Veinstein, 97–132. Paris: Peeters, 2008.

Kendall, Charles B. "Preparation of Alizarine Assistant and Its Action in Turkey-Red Dyeing." Massachusetts Institute of Technology, 1887.

Kent, Harold Winfield. *Treasury of Hawaiian Words in One Hundred and One Categories*. Honolulu: The Masonic Public Library of Hawaii, 1986.

Keyder, Çaglar, Y. Eyüp Özveren, and Donald Quataert. "Port-Cities in the Ottoman Empire: Some Theoretical and Historical Perspectives." *Review (Fernand Braudel Center)* 16, no. 4 (1993): 519–58.

Kiel, Eric George, and P.M. Heertjes. "I. The Metal Complexes of Alizarin Structure of the Calcium-Aluminium Lake of Alizarin." *Journal of the Society of Dyers and Colourists* 79, no. January (1963): 21–7.

Kiel, Eric George, and P.M. Heertjes. "Metal Complexes of Alizarin V. Investigations of Alizarin–Dyed Cotton Fabrics." *Journal of the Society of Dyers and Colourists* 81, no. 3 (1965): 98–102.

Kiel, Eric George. "Metaalcomplexen van alizarinerood." TU Delft, 1961.

Kirby, Jo, Marika Spring, and Catherine Higgitt. "The Technology of Eighteenth- and Nineteenth-Century Red Lake Pigments." *National Gallery Technical Bulletin* 28 (2007): 69–95.

Knecht, Edmund, Christopher Rawson, and Richard Loewenthal. *A Manual of Dyeing*. Vol. 1. London: Charles Griffin & Company, Limited, 1893.

Knecht, Edmund, Christopher Rawson, and Richard Loewenthal. *A Manual of Dyeing*. Vol. 2. London: Charles Griffin & Company, Limited, 1893.

Knoepfli, Albert. "Die Sulzersche Rotfarb und Kattun-Druckerei zu Aadorf." *Thurgauer Jahrbuch* 26 (1951): 24–38.

Koller, Theodor. *The Utilization of Waste Products*. 3rd edn. London: Scott, Greenwood & Son, 1918.

Koren, Zvi. "The Colors and Dyes on Ancient Textiles in Israel." In *Colors from Nature: Natural Colors in Ancient Times*, edited by Sorek C and E Ayalon, 15–31. Tel Aviv: Eretz Israel Museum, 1993.

Kusamitsu, Toshio. "British Industrialization and Design before the Great Exhibition." *Textile History* 12, no. 1 (1981): 77–95.

Lalande, F de. "Synthesis of Purpurin." *The Chemical News and Journal of Physical Science* 30, no. 779 (1874): 207.

Lalande, Joseph Jerôme le Français de. "L'art de faire le maroquin." In *Descriptions des arts et métiers, faites ou approuvées par Messieurs de l'académie royale des sciences*. Paris: Saillant et Nyon, 1761.

Le département Hauts-de-seine. "Archives de la Planète." Les collections du musée départemental Albert-Kahn. Accessed January 15, 2022. http://collections.albert-kahn.hauts-de-seine.fr/

Leigh, W.N. "On the Estimation of Alizarin in Dyed Cotton Fabrics, and on an Attempt to Ascertain the Composition of Turkey-Red and Other Alizarin Lakes." *Journal of the Society of Dyers and Colourists* 32, no. 8 (1916): 205–13.

Lemire, Beverly. "'A Good Stock of Cloaths': The Changing Market for Cotton Clothing in Britain, 1750–1800." *Textile History* 22, no. 2 (1991): 311–28.

Le Pileur d'Apligny. *Essai sur les moyens de perfectionner l'art de la teinture, et observations sur quelques matières qui y sont propres*. Paris: Saugrain, 1773.

Le Pileur d'Apligny. *L'art de la teinture des fils et étoffes de coton*. Paris: Servière, 1798.

Liles, J. N. *The Art and Craft of Natural Dyeing: Traditional Recipes for Modern Use*. Knoxville: University of Tennessee Press, 1990.

Long, Bridget. "Introduction." *Quilt Studies* 1 (1999).

Lopez, Judith. "The Transition from Natural Madder to Synthetic Alizarine in the American Textile Industry, 1870–1890." Iowa State University, 1989.

Lynde, J. *The Domestic Dyer, or Philosophy of Fast Colours: Being a Compilation from the Most Approved American and European Authors*. New York, NY, 1831.

MacDonald, Gerald J. "Spanish Textile and Clothing Nomenclature in -Án, -í, and Ín." *Hispanic Review* 44, no. 1 (April 7, 1976): 57–78.

MacFarlan, Duncan, John Forbes, John Lockhart, Robert Buchanan, John G Lorimer, John Smyth, Nathaniel Paterson, et al. "City of Glasgow and Suburban Parishes of Barony and Gorbals." In *The New Statistical Account of Scotland*, 6:101–241. Edinburgh: William Blackwood and Sons, 1845.

MacFarlane, Robert. *A Practical Treatise on Dyeing and Calico-Printing*. New York, NY: John Wiley, 1860.

MacKay, John. *Bleachfields, Printfields and Turkey Red*. Renton: The Carman Centre, 2011.

Macquer, P. *Dictionnaire portatif des arts et métiers: contenant en abrégé l'histoire, la description & la police des arts et métiers, des fabriques et manufactures de France et des pays étrangers*. Vol. 3. Dictionnaire portatif des arts et métiers: contenant en abrégé l'histoire, la description & la police des arts et métiers, des fabriques et manufactures de France et des pays étrangers. Yverdon: F.-B. de Félice, 1767.

Markley, Klare S. *Fatty Acids: Their Chemistry, Properties, Production, and Uses*. New York, NY: Interscience Publishers, Inc., 1964.

Masson, Paul. *Histoire du commerce français dans le Levant au XVIIIe siècle*. Paris: Librarie Hachette, 1911.

Mauro, Antonio., and Piero. Fiorenzani. "Sul 'Follone di Garibaldi' esistente a Prato." *Prato / Azienda Autonoma di Turismo di Prato*, 2012, 127–37.

Mazeas, Guillaume. "Recherches sur la cause physique de l'adhérence de la couleur rouge aux Toiles peintes qui nous viennent des cotes de Malabar & de Coromandel." In *Mémoires de mathématique et de physique, présentés à l'Académie royale des sciences, par divers sçavans, & lûs dans ses assemblées*, 4:1–32. Paris: Moutard, 1763.

Mazzolini-Trümpy, Hanruedi. "Von Fabrikanten, Teilhabern, Drogenhändlern und Versicherungsagenten in der Glarner Industrielandschaft des 19. Jahrhunderts." *Jahrbuch des Historischen Vereins des Kantons Glarus* 87 (2007).

Meyer, M., T. Huthwelker, C. N. Borca, K. Meßlinger, M. Bieber, R. H. Fink, and A. Späth. "In-Situ Spectroscopic Analysis of the Traditional Dyeing Pigment Turkey Red inside Textile Matrix." *Journal of Instrumentation* 13, no. 3 (2018).

Millar, John. "Mr. Millar's Statement of Claims, &c." *The Glasgow Mechanics' Magazine* 4, no. 114 (1826): 447.

Miller, John. "Reply to Mr. David Campbell's Statement of Claims to the Invention of the Process for Discharging Turkey Red. Inserted in the Glasgow Mechanics' Magazine, No. LXXXIX, Sep. 3d 1825." *The Glasgow Mechanics' Magazine* 4, no. 110 (1826): 381–3.

Miller, John. "Statement of Claims to the Invention of the Process for Discharging Turkey-Red." *The Glasgow Mechanics' Magazine*, no. 84 (1825): 410–14.

Miller, John. "Statement Relative to the Discharging Process of Turkey Red, by Means of Presses." *The Glasgow Mechanics' Magazine* and Annals of Philosophy 1, no. 29 (1824): 462–4.

Miller, Philip. *The Method of Cultivating Madder, as It Is Now Practised by the Dutch in Zealand.* London: Author, 1758.

Miquelon, Dale. *Dugard of Rouen: French Trade to Canada and the West Indies, 1729–1770.* Kingston, ON: McGill-Queen's University Press, 1978.

Mohanty, B.C., K.V. Chandramouli, and H.D. Naik. *Natural Dyeing Processes of India.* Ahmedabad: Calico Museum of Textiles, 1987.

Montgomery Ward and Company. "Catalogue No. 13 Spring and Summer 1875." Chicago, Il: Montgomery Ward & Co., 1875.

Moulherat, Christophe, Margareta Tengberg, Jérôme-F. Haquet, and Benoît Mille. "First Evidence of Cotton at Neolithic Mehrgarh, Pakistan: Analysis of Mineralized Fibres from a Copper Bead." *Journal of Archaeological Science* 29, no. 12 (2002): 1393–1401.

Mukund, Kanakalatha. "Indian Textile Industry in 17th and 18th Centuries: Structure, Organisation and Responses." *Economic and Political Weekly* 27, no. 38 (1992): 2057–65.

Müller-Jacobs, A. "Turkey-Red from Alizarine." *Scientific American* 47, no. 18 (1882): 280.

Musson, A.E., and Eric Robinson. "Science and Technology in the Industrial Revolution." Manchester: Manchester U.P., 1969.

Naughton, Frank C. "Production, Chemistry, and Commercial Applications of Various Chemicals from Castor Oil." *Journal of the American Oil Chemists' Society* 51, no. 3 (1974): 65–71.

Nenadic, Stana, and Sally Tuckett. *Colouring the Nation: The Turkey Red Printed Cotton Industry in Scotland c. 1840–1940.* Edinburgh: NMS Enterprises Limited, 2013.

Nenadic, Stana. "Selling Printed Cottons in Mid-Nineteenth-Century India: John Matheson of Glasgow and Scottish Turkey Red." *Enterprise and Society* 20, no. 2 (2019): 328–65.

Neyland, Robert S. "The Seagoing Vessels on Dilmun Seals." In *Underwater Archaeology Proceedings from the Society for Historical Archaeology Conference: Kingston, Jamaica 1992*, edited by Toni Carrell and Donald H Keith, 68–74. Pleasant Hill, CA: Society for Historical Archaeology, 1992.

Niederhäusern, F.-H. de. "Rapport sur le travail de M. F. Driessen: 'Etude sur le rouge turc, ancien procédé', et sur le contenu de deux plis cachetés Nos. 700 et 1276, déposés par le même auteur." *Bulletin de la Société Industrielle de Mulhouse* 72 (1902).

Nieto-Galan, Augusti. *Colouring Textiles : A History of Natural Dyestuffs in Industrial Europe*, 2001.

O'Connor, Deryn. "Four Aspects of Turkey Red: The Process and Early History." *Quilt Studies* 1 (1999).

O'Connor, Deryn. "Four Aspects of Turkey Red: The Turkey Red Industry: Export Cloths." *Quilt Studies* 1 (1999).

O'Neill, Charles, and A.A. Fesquet. *A Dictionary of Dyeing and Calico Printing*. Philadelphia: Henry Carey Baird, 1869.

O'Neill, Charles. "The Printing and Dyeing of Calico, Silk, and Woollen Fabrics." In *The Record of the International Exhibition, 1862*, edited by Robert Mallet. Glasgow: W. MacKenzie, 1862.

Oberholzer-Hofmann, Paul. *Die Rotfarb Uznach : hundert Jahre im Besitze der Familie Hofmann*. Uznach: Oberholzer, 1975.

Official Data Foundation. "Official Data Foundation," 2019. Accessed March 6, 2023. https://www.officialdata.org/about

Pallas, Simon Peter. "I. The Genuine Oriental Process for Giving to Cotton Yarn or Stuff the Fast or Ingrained Colour, Known by the Name of Turkey Red, as Practised at Astracan." *Philosophical Magazine Series 1* 1, no. 1 (1798): 4–11.

Pallas, Simon Peter. "II. Process for Dyeing the Adrianople or Turkey Red, as Practised at Astracan. Being a Supplement to His Former Publications on That Art." *The Philosophical Magazine* 25, no. 97 (1806): 8–9.

Parks, Lytle Raymond. "The Chemistry of Turkey-Red Dyeing." *Journal of Physical Chemistry* 35, no. 2 (1930): 488–510.

Peel, R. A. "Turkey Red Dyeing in Scotland Its Heyday and Decline." *Journal of the Society of Dyers and Colourists* 68, no. 12 (1952): 496–505.

Perkin, Arthur George. "Methods of Analysis Employed in the Manufacture of Alizarin." *The Journal of the Society of Dyers and Colourists* 13, no. 4 (1897): 81–7.

Perkin, William Henry. "The History of Alizarin and Allied Colouring Matters, and Their Production from Coal Tar." *Journal of the Society for Arts* 27, no. 1384 (1879): 572–608.

Perkin, William Henry. "XV. On Anthrapurpurin." *Journal of the Chemical Society*, no. 26 (1873): 425–33.

Perkin, William Henry. "XVI. On Artificial Alizarin." *Journal of the Chemical Society* 23 (1870): 133–43.

Perkin, William Henry. "XXXI. On the Formation of Anthrapurpurin." *Journal of the Chemical Society*, no. 29 (1876): 851–5.

Perkin, William Henry. Coloring Matter. GB18691948. United Kingdom: Office of the Commissioners for Patents, issued 1869.

Perkin, William Henry. Coloring Matter. GB18693318. United Kingdom: Office of the Commissioners for Patents, issued 1870.

Persoz, Jean-François. *Traité théorique et pratique de l'impression des tissus*. Vol. 1. Paris: Victor Masson, 1846.

Persoz, Jean-François. *Traité théorique et pratique de l'impression des tissus*. Vol. 3. Paris: Victor Masson, 1846.

Persoz, Jean-François. *Traité théorique et pratique de l'impression des tissus*. Vol. 4. Paris: Victor Masson, 1846.

Potukuchi, Swarnalatha. "The World of the Weaver in the Northern Coromandel, 1750–1850." University of Hyderabad, 1991.

Quinton, Rebecca. "Turkey Red and the Slave Economy." Legacies of Slavery in Glasgow Museums and Collections, 2019. Accessed March 6, 2023. https://glasgowmuseumsslavery.co.uk/2019/05/09/turkey-red-and-the-slave-economy/

Radcliffe, L.G., and S. Medofski. "The Sulphonation of Fixed Oils." *Journal of the Society of Dyers and Colourists* 34, no. 2 (1918): 22–35.

Rauch, John, and William. Sherman. "John Rauch's Receipts on Dyeing Cotton & Woolen." Receipts on Dyeing. Plympton MA, 1816. Accessed March 6, 2023. https://digital.clarkart.edu/digital/collection/p16245coll1/id/100348

Rauch, John. *John Rauch's Receipts on Dyeing, in a Series of Letters to a Friend. Containing Correct and Exact Copies of All His Best Receipts on Dyeing Cotton and Woollen Goods; Obtained and Improved by Him, during Twelve Years Practice, at Different Manufactories in Switzerland, France, Germany and America. Also, a true Description of His Invented Substitute for Woad, Being a Cheap and Preferable Material, and the Produce of this Country.* New York, NY: John Sterly Ermantinger, 1815.

Raveux, Olivier. "Spaces and Technologies in the Cotton Industry in the Seventeenth and Eighteenth Centuries: The Example of Printed Calicoes in Marseilles." *Textile History* 36, no. 2 (2005): 131–45.

Raveux, Olivier. "The Orient and the Dawn of Western Industrialization: Armenian Calico Printers from Constantinople in Marseilles (1669–1686)." In *Goods from the East, 1600–1800*, edited by Maxine Berg, Felicia Gottmann, Hanna Hodacs, and Chris Nierstrasz, 77–91. Basingstoke: Palgrave MacMillan, 2015.

Reinking, Karl, and Sabri Atayolu. "Zur Entstehung und Frühgeschichte des Türkischrots." *Melliand Textilberichte* 18 (1937): 382–84, 459–60, 532.

Richter, Derek. "LXXXII. Vital Staining of Bones with Madder." *Biochemical Journal* 31, no. 4 (1937): 591–5.

Riello, Giorgio. *Cotton: The Fabric That Made the Modern World.* Cambridge: Cambridge University Press, 2013.

Robinson, Stuart. *A History of Dyed Textiles.* Cambridge, MA: MIT Press, 1969.

Robinson, Stuart. *A History of Printed Textiles.* Cambridge, MA: MIT Press, 1969.

Rouffaer, Gerrit Peter, and H H Juynboll. *De batik-kunst in Nederlandsch-Indië en haar geschiedenis.* [Publicaties van 's Rijks ethnographisch museum. ser. II, no. 1]. Utrecht: A. Oosthoek, 1914.

Rowe, F.M. "The Life and Work of Sir William Henry Perkin." *Journal of the Society of Dyers and Colourists* 54, no. 12 (1938): 551–62.

Russell, Frank. "Personal Conversation." Malden, MA: Unpublished, 2020.

Sandberg, Gösta. *The Red Dyes.* Translated by Edith M Matterson. Asheville, NC: Lark Books, 1997.

Sansone, Antonio. "Alizarin-Red and Turkey-Red Dyeing and Printing on Cotton." *Journal of the Society of Dyers and Colourists* 1, no. 8 (1885): 203–11.

Santis, D. De, and M. Moresi. "Production of Alizarin Extracts from Rubia Tinctorum and Assessment of Their Dyeing Properties." *Industrial Crops and Products* 26, no. 2 (2007): 151–62.

Schaefer, Gustav. "The Cultivation of Madder." *Ciba Review* 39, no. May (1941): 1398–1406.

Schaefer, Gustav. "The History of Turkey Red Dyeing." *Ciba Review* 39, no. May (1941): 1407–16.

Schmitt, J. M. "Relations between England and the Mulhouse Textile Industry in the Nineteenth Century" 4969, no. September (1986).

Schmitt, Marco, Sven Boras, Aiyen Tjoa, Toshihiro Watanabe, and Steven Jansen. "Aluminium Accumulation and Intra-Tree Distribution Patterns in Three Arbor Aluminosa (Symplocos) Species from Central Sulawesi." *PLOS ONE* 11, no. 2 (February 12, 2016).

Schunck, Edward. "XX. On the Colouring Matters of Madder." *Quarterly Journal of the Chemical Society of London*, no. 12 (1860): 198–221.

Schutzenberger, Paul. *Traite des Matières Colorantes*. Vol. 2. Paris: Victor Masson et Fils, 1867.

Schweppe, Helmut, and John Winter. "Madder and Alizarin." In *Artists' Pigments: A Handbook of Their History and Characteristics*, edited by Elisabeth West, 109–42. Washington DC: National Gallery of Art, 1998.

Scoggin, Andrew. "A Piece of Red Calico." *Scribners Monthly*. New York, NY, February 1876.

Scott, Ashley, Robert C Power, Victoria Altmann-Wendling, Michel Artzy, Mario A S Martin, Stefanie Eisenmann, Richard Hagan, et al. "Exotic Foods Reveal Contact between South Asia and the Near East during the Second Millennium BCE." *Proceedings of the National Academy of Sciences* 118, no. 2 (2021).

Smith, Michael Stephen. *The Emergence of Modern Business Enterprise in France, 1800–1930*. Harvard Studies in Business History. Cambridge, MA and London, England: Harvard University Press, 2006.

Smith, Tina Fenwick. "Four Aspects of Turkey Red," 1999.

Soubayrol, Patrick, Gilbert Dana, and Pascal P. Man. "Aluminium-27 Solid-State NMR Study of Aluminium Coordination Complexes of Alizarin." *Magnetic Resonance in Chemistry* 34, no. 8 (1996): 638–45.

Soubayrol, Patrick, Gilbert Dana, G Bolbach, and J.C. Tabet. "Spectrométrie de masse et chromatographie liquides des laques complexes des teintures à l'alizarine." *Analusis* 24, no. 7 (1996): 34–36.

Soubayrol, Patrick. "Préparation et étude structurale des complexes formes entre l'aluminium et l'alizarine. Importance de la nature du solvant et de la base utilises sur le degré de condensation de l'aluminium et l'agencement moléculaire." Université Pierre et Marie Curie, 1996.

St Clair, Kassia. *The Golden Thread: How Fabric Changed History*. New York and London: Liveright Publishing Corporation, 2018.

Stewart, Gordon. "Parish of Bonhil." In *The Statistical Account of Scotland*, edited by John Sinclair, 442–53. Edinburgh: William Creech, 1792.

Stirling, John. "History of Colour Printing in the United Kingdom." *Journal of the Society of Dyers and Colourists*, no. February (1903): 36–40.

Storey, Joyce. "Just New From the Mills—Printed Cottons in Victorian America." *Textile History* 15, no. 2 (1984): 246–48.

Storey, Joyce. "Turkey Red Prints." *Surface Design Journal* 29, no. 4 (1996).

Storey, Joyce. *The Thames and Hudson Manual of Dyes and Fabrics*. London: Thames and Hudson, 1978.

Straugh, Mr. "Recipe from Mr Straugh." Paisley: Paisley Central Library, 1908.

Swain, Margaret H. "Turkey Red." *The Scots Magazine*, no. March (1965): 536–41.

Tannahill, John. "Turkey Red Dyeing." Paisley: Paisley Central Library, 1912.

Tarrant, Naomi E A. "The Turkey Red Dyeing Industry in the Vale of Leven." In *Scottish Textile History*, edited by John Butt and Kenneth G Ponting, 37–47. Aberdeen University Press, 1987.

The Editors of Encyclopaedia Britannica. "Zhao Rukuo." *Encyclopaedia Britannica*, n.d. Accessed March 6, 2023. https://www.britannica.com/biography/Zhao-Rukuo

The National Archives. "The National Archives Currency Converter," n.d. Accessed March 6, 2023. https://www.nationalarchives.gov.uk/currency-converter/

Thomson, John. "I. New Theory of Dyeing Turkey Red." In *Annals of Philosophy*, Vol. 8. London, 1816.

Thorsen, Linda Jean. "The Merchants and the Dyers : The Rise of a Dyeing Service Industry in Massachusetts and New." In *Crosscurrents: Land, Labor, and the Port. Textile Society of America's 15th Biennial Symposium*, 490–9. Savannah, GA: Textile Society of America, 2016.

Trask, R. Hugh. "Sulfonation and Sulfation of Oils." *Journal of the American Oil Chemists' Society* 33, no. 11 (1956): 568–71.

Travis, Anthony S. "Between Broken Root and Artificial Alizarin: Textile Arts and Manufactures of Madder." *History and Technology: An International Journal* 12, no. 1 (1994): 1–22.

Travis, Anthony S. *The Rainbow Makers: The Origins of the Synthetic Dyestuffs Industry in Western Europe.* London: Associated University Presses, 1993.

Trotman, E.R. *Dyeing and Chemical Technology of Textile Fibres.* 5th edn. London: Charles Griffin & Company, Limited, 1975.

Tuckett, Sally, and Stana Nenadic. "Colouring the Nation: A New In-Depth Study of the Turkey Red Pattern Books in the National Museums Scotland." *Textile History* 43, no. 2 (2012): 161–82.

Ure, Andrew. "Description of the Great Bandana Gallery in the Turkey-Red Factory of Messrs. Monteith & Co. at Glasgow." *The Glasgow Mechanics' Magazine* 1, no. 1 (1824).

Ure, Andrew. *A Dictionary of Arts, Manufactures, and Mines.* 3rd edn. Vol. 2. New York, NY: D. Appleton & Company, 1844.

Venkataraman, K. *The Chemistry of Synthetic Dyes.* Vol. 1. New York, NY: Academic Press Inc., 1952.

Verbong, Geert P.J. "De ontwikkeling van het stakingsrecht in Nederland." *Jaarboek voor de geschiedenis van bedrijf en techniek* 3 (1987): 183–204.

Verbong, Geert P.J. "Technische innovaties in de katoendrukkerij en -ververij in Nederland 1835–1920." TU Eindhoven, 1988.

Verbong, Geert P.J. "Turksrood." In *Geschiedenis van de techniek in Nederland. De wording van een moderne samenleving 1800–1890. Deel V. Techniek, beroep en praktijk*, edited by H.W. Lintsen. Zutphen: Walburg Pers, 1994.

Vitalis, Jean-Baptiste. "Mémoire sur la nature la fiente de mouton, et sur son usage dans la teinture du coton en rouge di des Indes, ou d'Andrinople." *Journal de physique, de chimie, d'histoire naturelle et des arts* 66, no. February (1808): 153–61.

Vitalis, Jean-Baptiste. *Cours élémentaire de teinture sur laine, soie, lin, chanvre et coton, et sur l'art d'imprimer les toiles.* Paris: Galerie Bossange Pere, 1823.

Vitalis, Jean-Baptiste. *Essai sur l'origine et les progrès de l'art de la teinture en France, et particulièrement de l'art de teindre le coton en rouge dit des Indes. Lu à la Société de commerce de Rouen.* de l'imp. de P. Periaux, 1808.

Wadsworth, Alfred P, and J De Lacy Mann. *The Cotton Trade and Industrial Lancashire, 1600–1780.* Manchester University Press, 1931.

Walker, N. "The Life and Influence of Professor J.J. Hummel." *Journal of the Society of Dyers and Colourists*, no. January (1939): 14–30.

Washington, L.A. "On Dyeing Turkey Red." *Archives of Useful Knowledge, a Work Devoted to Commerce, Manufactures, Rural and Domestic Economy, Agriculture, and the Useful Arts (1810–1813)* 2, no. 1 (1811): 50–3.

Welham, R. D. "The Early History of the Synthetic Dye Industry." *Journal of the Society of Dyers and Colourists* 79, no. 5 (1963): 181–5.

Welters, L, and M T Ordoñez. *Down by the Old Mill Stream: Quilts in Rhode Island.* Kent, OH: Kent State University Press, 2000.

Wertz, Julie H, Pik Leung Tang, Anita Quye, and David J France. "Characterisation of Oil and Aluminium Complex on Replica and Historical 19th c. Turkey Red Textiles by Non-Destructive Diffuse Reflectance FTIR Spectroscopy." *Spectrochimica Acta Part A: Molecular and Biomolecular Spectroscopy* 204 (2018): 267–75.

Wertz, Julie H. "Turkey Red Dyeing in Late-19th Century Glasgow: Interpreting the Historical Process through Re-Creation and Chemical Analysis for Heritage Research and Conservation." University of Glasgow, 2017.

Wertz, Julie H., Anita Quye, and David France. "Turkey Red Prints: Identification of Lead Chromate, Prussian Blue and Logwood on Turkey Red Calico." *Conservar Património* 31 (2019): 31–9.

Wertz, Julie H., David J. France, and Anita Quye. "Spectroscopic Analysis of Turkey Red Oil Samples as a Basis for Understanding Historical Dyed Textiles." *Coloration Technology* 134, no. 5 (2018): 319–26.

Wilder, Laura Ingalls. *Little House in the Big Woods*. New York, NY: Harper & Row, 1932.

William Sterling & Sons. "Private Ledger." Glasgow, n.d.

Wilson, Arthur. "Turkey-Red Oil Part II." *Journal of the Society of Chemical Industry*, 1892, 495–6.

Wilson, John. *As Essay on Light and Colours, and What Colouring Matters Are That Dye Cotton and Linen*. Manchester: J. Harrop, 1786.

Wood, Frances Gilchrist. "Turkey Red." In *The Best Short Stories of 1920*, edited by Edward J O'Brien. Boston: Small, Maynard & Company, 1920.

Wunderlich, Christian-Heinrich, and G Bergerhoff. "Crystal Structure of di-μ-oxo-bis(bis(1,2-dihydroxyanthrachinonato)aluminium-(aquabis(dimethylformamide)calcium)), $Al_2O_4(C1_4H_6O_4)_4Ca_2$ $(HCON(CH_3)_2)_4(H_2O)_2$, (alizarine complex)." *Zeitschrift Für Kristallographie—Crystalline Materials* 207, no. 1 (1993): 185–8.

Young, J. Wallace. "XI.—On Artificial Alizarine." *Proceedings of the Philosophical Society of Glasgow* 7 (1871): 316–18.

Zhuang, Guanzheng, Silvia Pedetti, Yoan Bourlier, Philippe Jonnard, Christophe Méthivier, Philippe Walter, Claire-Marie Pradier, and Maguy Jaber. "New Insights into the Structure and Degradation of Alizarin Lake Pigments: Input of the Surface Study Approach." *The Journal of Physical Chemistry C* 124, no. 23 (June 11, 2020): 12370–80.

Index